PHOTOCATALYSTS - APPLICATIONS AND ATTRIBUTES

Edited by **Sher Bahadar Khan**
and **Kalsoom Akhtar**

Photocatalysts - Applications and Attributes
http://dx.doi.org/10.5772/intechopen.75848
Edited by Sher Bahadar Khan and Kalsoom Akhtar

Contributors

Muhammad Saqib Ishaq, Zobia Afsheen, Amjad Khan, Endang Tri Wahyuni, Nurul Hidayat Aprilita, Florica Manea, Corina Orha, John Moma, Jeffrey Baloyi, Linh Nguyen, Kwangsoo No, Seungbum Hong, Ayman Zaki, Waleed El Rouby, Vladimir Alonso Escobar Barrios, Nancy Cervantes-Rincón, Alma Jasso-Salcedo, Dalia Verónica Sánchez Rodríguez, Gustavo Colpani, Cíntia Soares, Natan Padoin, Humberto Gracher Riella, Micheli Zanetti, Rubieli Carla Frezza Zeferino, Luciano Luiz Silva, Josiane Maria Muneron Mello, Márcio Fiori

First published in London, United Kingdom, 2019 by IntechOpen
IntechOpen is the global imprint of INTECHOPEN LIMITED, registered in England and Wales, registration number: 11086078, The Shard, 25th floor, 32 London Bridge Street
London, SE19SG – United Kingdom
Printed in Croatia

British Library Cataloguing-in-Publication Data
A catalogue record for this book is available from the British Library

Additional hard copies can be obtained from orders@intechopen.com

Photocatalysts - Applications and Attributes, Edited by Sher Bahadar Khan and Kalsoom Akhtar
p. cm.
Print ISBN 978-1-78985-475-6
Online ISBN 978-1-78985-476-3

For EU product safety concerns:
IN TECH d.o.o., Prolaz Marije Krucifikse Kozulić 3, 51000 Rijeka, Croatia,
info@intechopen.com or visit our website at intechopen.com.

Meet the editors

Dr. Sher Bahadar Khan received his Ph.D in chemistry from HEJ, Karachi University, Pakistan. After completion of his Ph.D, he started his post-doctoral career in nano-chemistry and nanotechnology and continued to work as a post-doctoral research fellow until February 2010 at Yonsei University, South Korea. In March 2010, he joined the Center for Advanced Materials and Nano-engineering, Department of Chemistry, Najran University as an Assistant Professor and continued his work until August 2011. He joined the Chemistry Department, King Abdulaziz University, Jeddah, Saudi Arabia as an Assistant Professor in September 2011. Currently he is a full Professor and performs research in nanochemistry and nanotechnology, which comprises drug delivery, solar cell, development of active photo-catalysts, and fabrication of perceptive bio-sensors and chemi-sensors using metal oxide nanomaterials and their green environmental nanohybrids. He is the author of six books, six patents, and 270 research papers with more than 800 impact factor, 4000 citations, and 3a 5 h-index.

Dr. Kalsoom Akhtar received her Ph.D from the Chemistry Department, Ewha Womans University, Seoul, Korea. Dr. K. Akhtar is an Assistant Professor at the Chemistry Department, King Abdulaziz University and performs research in organic and nano-chemistry, which comprises photo-catalysts, organic synthesis, and metal oxide nanomaterials. She is the author of two books and 65 research papers.

Contents

Preface

Photocatalysts are based on nanomaterials and their applications are extremely broad. This is a constantly growing research area and it has become an important research topic in the recent years. Solar light photocatalysis is currently one of the most blossoming interdisciplinary fields of science. Research in the development of efficient photocatalytic materials has seen significant progress in the last two decades with a large number of research papers published every year. Improvements in the performance of photocatalytic materials have been largely correlated with advances in nanotechnology. As photocatalysts have diverse applications, the interest on photocatalysis has accelerated and has attracted worldwide interest. Photocatalysts have potential to be used in photodegradation of several contaminants to solve environmental problems. Other important applications of photocatalysts are the development of self-cleaning products, air purification, CO_2 conversion to hydrocarbon fuels as well as the production of highly effective fuels such as hydrogen.

Hence, photocatalysis is a subject of key interest. Therefore, it was planned to present recent progress in photocatalysis in the form of a book to the scientific community. There are eight chapters in this book. In the first chapter, different methods for disinfection such as chlorination, zone, ultraviolet light, chloramines, potassium permanganate, photocatalytic disinfection, nanofiltration, and chlorine dioxide are discussed. The second chapter is related to the selective degradations using a titanium-based photocatalyst while in chapter three, some of the most recent works that have employed the doping, decoration, and structural modification of TiO_2 particles for applications in photocatalysis have been reviewed. Additionally, the effectiveness of these dopants and/or modifiers in enhancing TiO_2 photoactivity as well as some perspective on the future of TiO_2 photocatalysis are also discussed. Chapter 4 presents modified materials such as alternatives for conventional photocatalysts as titanium dioxide. Discussion about silver/graphene nanoparticle-modified zinc oxide for degradation of pollutants such as triclosan or bisphenol A, both considered as endocrine disruptors, which affect the hormonal development of humans, is presented. In addition, bismuth oxychloride has gained attention during the last 5 years for photocatalysis. In accordance, the obtained results for phenol photodegradation, using such oxychloride, are also presented. In the chapter, characterization of the photocatalyst is reported along with the proposal for mechanisms of action for the modified ZnO photocatalyst and bismuth oxychloride. In Chapter 5, structural modifications in semiconductors have been proposed to enhance the photocatalytic activity, such as doping processes with elements that are capable of generating superficial defects that capture the formed electrons, avoiding the recombination or increasing the density of –OH groups or water molecules on the surface of the catalyst, which can enhance the formation of hydroxyl radicals. Therefore, this brief review is proposed to show the role of lanthanides in TiO_2 doping and the synthesis method applied, as well as the results discussed in the literature. Chapter 6 presents the role of various carbon forms, i.e., activated carbon and carbon nanotubes/nanofibers as support for TiO_2 in drinking water treatment is discussed. Also, the chapter discusses how TiO_2 is supported on zeolite to act bi-functionally as a sorbent/photocatalyst for drinking water treatment. The main contaminants of natural organic matter (NOM), arsenic species, and nitrogen compounds from drinking water sources by type of groundwater and surface water can be removed/degraded by sorption/photocatalysis using TiO_2 supported onto carbon and/or zeolite. TiO_2 supported on powdered activated carbon (PAC-TiO_2), granular activated carbon (GAC-TiO_2), and zeolite (Z-TiO_2), namely supported TiO_2, were synthesized through sol-gel method and TiO_2 and multi-wall carbon nanotubes/carbon nanofibers dispersed within epoxy matrix (CNT-TiO_2-Epoxy; CNF-TiO_2-Epoxy), namely TiO_2 composite, were obtained through two-roll mill method. Kinetics study results using specific mathematic models allowed the elucidation of some mechanistic aspects for sorption and photocatalysis for the application in drinking water. The intercalation of the carbon and zeolite supported TiO_2 layers into a filtering system that allows to develop a self-cleaning filtering system in drinking water. Chapter 7 presents a review of the role of HAp in the TiO_2/Hydroxyapatite composite including the adsorption ability of contaminations and the promoted impacts of the HAp component. Chapter 8 presents the study of the TiO_2 photocatalyst for photoreduction of several reducible chemicals. The photocatalytic reduction of several toxic metal ions, including Ag(I), Cu(II), Cr(VI), Hg(II), and U(VI) in the presence of TiO_2 , in order to decrease their toxicity, is described. Photodeposition of the noble metals, such as Ag(I), Au(III), Pt(IV), and Pd(II) for doping purposes by photocatalytic reduction over TiO_2 is also addressed. Conversion of the greenhouse gas of CO_2 into useful hydrocarbons and methanol by photocatalytic reduction using TiO_2 photocatalyst is highlighted. Several operating parameters in photoreduction processes (photocatalyst dose, time of irradiation, pH of the solution, and the initial concentration of the substrates (the reducible chemicals)) are also reviewed.

Dr. Sher Bahadar Khan
King Abdulaziz University
Jeddah, Saudi Arabia

Dr. Kalsoom Akhtar
Ewha Womans University
Seoul, South Korea

Section 1

Photocatalysts for Water Treatment

Chapter 1

Disinfection Methods

Muhammad Saqib Ishaq, Zobia Afsheen,
Amjad Khan and Amjad Khan

Additional information is available at the end of the chapter

http://dx.doi.org/10.5772/intechopen.80999

Abstract

Water must be made safe to drink, and an important step in ensuring water safety is disinfection. Disinfectants are added to water to kill disease-causing microorganisms. Ground water sources can be disinfected by "The Water Treatment Rule," which requires public water systems for disinfection. Chlorination, ozone, ultraviolet light, and chloramines are primary methods for disinfection. However, potassium permanganate, photocatalytic disinfection, nanofiltration, and chlorine dioxide can also be used. Organic material is naturally present in water. Certain forms of chlorine can react with these organic materials and result in the formation of harmful by-products; the U.S. Environmental Protection Agency has anticipated maximum levels for these contaminants.

Keywords: chlorination, chloramines, ozone, ultraviolet light, photocatalytic disinfection

1. Introduction

Killing, removal, or deactivation of harmful microorganisms can be referred to as disinfection. Destruction or deactivation of pathogenic microorganisms results in stopping their reproduction and growth. People may fall ill by consuming the contaminated water containing the pathogenic microorganisms. Disinfection and sterilization are interrelated processes, but sterilization kills all the harmful and harmless microorganisms. Hence, disinfection is a more appropriate process.

2. Methods of disinfections

2.1. Chlorine Gas

Chlorine is a greenish-yellow gas. By providing high pressure, the gas becomes liquid. It is toxic. Chlorine gas is mostly used as a water disinfectant. Introducing chlorine to water plays a very effective role for removing almost all pathogenic microorganisms. It can be used both as a primary and a secondary disinfectant. The gas is not applicable to be used in household system as it is very dangerous. It is lethal at concentrations as low as 0.1% air by volume [1].

2.1.1. Advantages

- Chlorination is a cheaper source than UV or ozone disinfection methods used to treat water.
- It is very effective against a wide range of pathogenic microorganisms.
- Dosing rates are controlled easily as they are flexible.
- The chlorine residuals left in the wastewater effluent can make the disinfection process longer even after initial treatment. They can be further used to evaluate the effectiveness [2].

2.1.2. Limitations

Although chlorine gas is used in large-scale water distribution treatment plants and networks as a best method for treating water, still it have various limitations. These limitations might affect the applicability to a point of use (POU) treatment system. Objections against chlorination are because of the esthetic, logistic, and health-related concerns.

Regarding esthetic level, chlorination might be rejected as it imparts bad tastes and odors to the water. The developed countries might teach their people about the good impacts of chlorination; however, less-developed countries lack this ability.

Limitations in using chlorine gas in a household context might include the distribution, procurement/manufacturing, dosing of chlorine, and accurate handling. The health hazards caused by chlorine are not only confined to its volatile nature. A great concern might be the byproducts and incompletely oxidized compounds present in chlorinated water that increases its toxicity. The most notorious byproducts of chlorination are chloro-organics and trihalomethane (THMs). Humic and fulvic acids are present in the water. When chlorine reacts with these acids, trihalomethane are formed. It has been identified in many studies that some of these chloro-organics are mutagens, toxins, or carcinogens. The well-known THM chloroform is an animal carcinogen. Some guidelines have been set by USEPA (United States Environmental Protection Agency) that THMs should not be greater than 0.10 mg/l. The high concentrations of THMs will lead to health complications [1].

2.1.3. Process

Chlorine readily combines with all the water components, i.e., chemicals, small animals, microorganisms, plant material, odors, colors, and tastes. Sufficient quantity of chlorine is necessary to meet with the demand of chlorine in the water and provide residual disinfection.

Residual (free) chlorine is termed as the one that does not combine with other water components. The point at which free chlorine is available for continuous disinfection is termed as the breakpoint. The system at which free chlorine is supplied at a concentration of 0.3–0.5 mg/l is an ideal system. Simple test kits, most commonly the DPD (N,N-diethyl-p-phenylenediamine) colorimetric test kits are available for testing breakpoint and chlorine residual in private systems. The kit must test the amount of free chlorine, not total chlorine [3].

2.1.4. Equipment

This consists of 908 kg (2000 lb) or 68 kg (150 lb) containers, weigh scales, chlorinator, injectors, switchover modules, vacuum lines, booster pumps, solution lines, diffusers, and a flow meter. Safety requirements are passive ventilation, mechanical ventilation, warning alarms and devices, showers, panic hardware for doors, and eye washes. A separate air tight room for chlorination equipment is mandatory. For disinfection process, 100% of chlorine gas is available [4].

2.1.5. Chemical

Chlorine can be used as liquid or gas form. It is a very strong, oxidizing agent. Both the forms (liquid and gas) can be stored and used from gas cylinders under pressure. The chlorine cylinders can be 150 pounds. Small drinking water systems commonly use cylinders of 150 pounds.

Hypochlorous and hypochlorite ions are formed when chlorine is mixed with water. The hypochlorous ion is a better disinfectant which is formed in greater concentration at low pH concentrations. The hypochlorite and hypochlorous ions will be present in equal concentrations at pH 7.3. At pH above 8.3, the hypochlorite ion predominates which is not a better disinfectant. So, better disinfection is achieved at low pH. To avoid the formation of trihalomethanes and haloacetic acids, chlorination should be applied after treatment [5] (**Figure 1**).

2.2. Chlorination (sodium hypochlorite solution)

Sodium hypochlorite is used as a bleaching agent, mostly to bleach papers or textiles, and as a disinfectant in solution. The solution generally contains 10–15% of the available chlorine, but rapidly loses its force in storing process. Regular controlled environment is needed as the solution is affected greatly by the pH, light, heat, and heavy metals [6].

2.2.1. Advantages

- Sodium hypochlorite can also be used as a disinfectant.
- Both sodium hypochlorite and chlorine gas shows similar disinfection effectiveness.

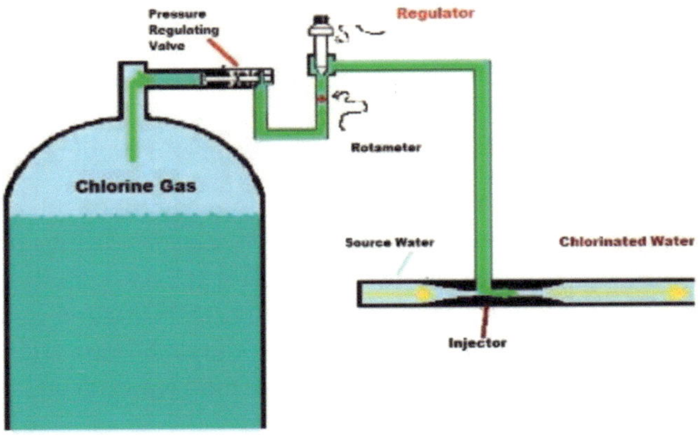

Figure 1. Chlorination by gas method.

- As compared to chlorine gas, sodium hypochlorite disinfection reduces the hazards in storing and handling.
- No hazardous chemicals are used in onsite generation. Only softened water and high grade salt (NaCl) is used.
- As compared to the standard supplied solution (14% concentration), sodium hypochloride (NaOCl) solutions are less hazardous (1% concentration) and less concentrated while generating onsite production [7]

2.2.2. *Limitations*

NaOCl can be commercially supplied or generated on-site, the latter being the safer of the two methods for handling reasons. In on-site generation, salt is dissolved with softened water to form a concentrated brine solution that is subsequently diluted and passed through an electrolytic cell to form sodium hypochlorite. Hydrogen is also produced during electrolysis, and it needs to be vented because of its explosive nature [7].

2.2.3. *Process*

These dosing systems are mostly simple, but there might be a concern regarding the design. The design might influence the control of gas release from the bulk hypochlorite in dosing pumps and piping and scale formation. Gasification (mostly produces oxygen) can lead to vapor or gas bubbles' formation, specifically if sodium hypochlorite is below the atmospheric pressure, which leads to gas locking of the suction line in a diaphragm. Pumps should therefore be provided with flooded suction. Tanks must be properly vented out of all structures to the atmosphere.

Most commonly available dosing systems use diaphragm metering pumps. The pump action can lead to the development of vacuum. The vacuum causes the vaporization of the dissolved gases in the sodium hypochlorite, resulting in the pump losing its prime and a lower applied chlorine dose.

Consequently, to aid in the prevention of gasification, the dosing arrangements must have a positive head on the pump suction (with pump inlet always below the minimum tank liquid level). In addition, piping system configurations which will trap sodium NaOCl between two closed isolation valves or check valves should be avoided.

A calibration cylinder, a pulsation damper, a pressure relief valve, and a loading valve are the main components of a dosing system. Automatic auto-degas valves systems are also provided by some dosing pump suppliers. NaOCl is dosed either through a spreader bar submerged within an open channel or through an injection fitting (pressurized pipes). The pulsation damper and the dosing pump should be placed close to each other. Pulsation damping also helps in improving dispersion. A loading valve should also be provided in systems where the back pressure at the pump delivery side is not sufficient (<0.7–1.0 Bar), till a suction demand valve is installed on the suction side.

For protecting the diaphragm from rupturing, a PRV (pressure relief valve) should also be provided on the delivery side of the pump. PRV's operation must be detected and alarmed: e.g., the outlet of the valve could be directed to a small "catch-pot," equipped with a float switch. All the systems or pumps which are shut down should contain methods to relieve any pressure buildup.

As sodium hydroxide (NaOH) is used in its manufacturing, the pH of NaOCl is high. Extreme care is needed when using hard waters (or waters with CO_2 present), as the highly alkaline product can lead to lower flow rates, reduced pipe diameter, reduce pump capacities, and scale formation at dosing points [8].

2.2.4. Equipment

NaOCl solution is a corrosive liquid with a high pH, i.e., 12. So, general precautionary measures for dealing with corrosive materials should be used such as avoiding metal contact, including stainless steel. These solutions might contain chlorate. Due do product degradation, chlorate can be formed during the processes of manufacturing and storage of sodium hypochlorite. The formation of chlorate and the degradation of NaOCl are directly related to each other. By reducing the degradation of NaOCl, the formation of chlorate can be minimized by avoiding high temperatures, reducing light exposure, and through limiting storage time. Spill containment should be provided for the NaOCl storage tanks. Typical spill containment structures include no uncontrolled floor drains, containment for the entire contents of the largest tank (along with the freeboard for rainfall/fire sprinklers), and separate containment areas for each incompatible chemical [9].

2.2.5. Chemicals

NaOCl solution (or liquid bleach) is a solution with 5–15% concentration of chlorine. It is used as bleaching and cleaning agent. It is also used extensively as water disinfectant, but it may not be the most economical solution as it is more expensive than gas. Being liquid, it can be handled easily than gas or calcium hypochlorite, but it is limited by its lack of stability and corrosive nature. It can be produced easily. Continuous supply of salt and electricity is needed for onsite generation of liquid bleach. Liquid bleach has a better POU application because of its availability and relative manageability [1] (**Figure 2**).

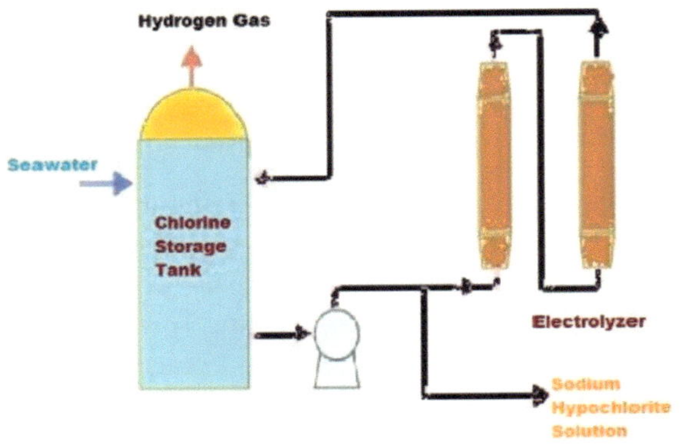

Figure 2. Chlorination by chlorine liquid method.

2.3. Chlorination (solid calcium hypochlorite)

$Ca(OCl)_2$ (calcium hypochlorite) is an essential solid that can be used in replacement of NaOCl (liquid). As a disinfectant, it has similarity with NaOCl but it is much safer to handle. Almost 70% of chlorine is available in commercial grades of $Ca(OCl)_2$. It has applications in both waste water and drinking water [7].

2.3.1. *Advantages*

- Being solid, $Ca(OCl)_2$ is more safer than chlorine gas and NaOCl.
- It even has excellent stability when stored in dry place, maintaining its potency well over time [7].

2.3.2. *Limitation*

Contamination or improper use of $Ca(OCl)_2$ may lead to explosion, fire, or the release of gases (toxic gases). Calcium hypochlorite should not be allowed to contact any foreign matter (including other water treatment products). If $Ca(OCl)_2$ is exposed to even very small amounts of water, it can react violently to produce toxic gases, heat, and spatter. Product should be added to water instead of adding water into the product. Exposure to heat can cause $Ca(OCl)_2$ to decompose rapidly, which may lead to explosion, intense fire, and the release of toxic gases. Dry, cool, well-ventilated area is needed for storing the product. $Ca(OCl)_2$ is used as a strong oxidizing agent. It increases the intensity of fire. $Ca(OCl)_2$ must be kept away heat, i.e., flame, heat, and any kind of burning materials [7].

2.3.3. *Process*

The calcium hypochlorite chlorinator contains a cylindrical polyvinyl chloride (PVC) tank with a height of 0.6–1.2 m and a diameter of 230–610 mm. A sieve plate is present containing holes that support the 80 mm diameter $Ca(OCl)_2$ tablets. Tablet chlorinator systems can

http://dx.doi.org/10.5772/intechopen.80999

usually provide between 1 and 295 kg of chlorine per day. At the bottom, a side stream is piped into the chlorinator. The flow arises from the holes in the sieve plate that results in eroding the last layer of tablets. The amount of water entering the chlorinator is proportional to the rate at which the tablets erode. The rate of chlorine dosage can be calculated by controlling the water flow through the chlorinator. To meet the operational requirement, the chlorinator effluent is returned to the main stream, providing the desired level of available chlorine.

Variation in the dose and the contact time can be done to calculate the required disinfection. Different factors affect the chlorine dosage, i.e., wastewater characteristics, the demand of chlorine and discharge requirements. Mostly, the dose ranges from 5 to 20 mg/L. Different factors account for optimum disinfection that might include temperature, alkalinity, and nitrogen content. The pH of wastewater might affect the distribution of chlorine between hypochlorous acid and hypochlorite. A lower pH favors the hypochlorous acid: a better disinfectant. Higher concentrations of hypochlorous acid will lead to the formation of hazardous chlorine gas [2].

2.3.4. Equipment

$Ca(OCl)_2$ can be added to the wastewater by two ways, i.e.,

1. Either by mixing calcium hypochlorite powder in a mixing device and then injecting it into the wastewater stream;
2. By immersing chlorine tablets in the wastewater using a tablet chlorinator [2] (**Figure 3**).

2.4. Chloramines

Chloramines are formed by reacting ammonia with free chlorine. They play an important role in providing residual protection in the distribution system. They are very stable. In comparison to chlorine, fewer halogenated by-products are formed [10].

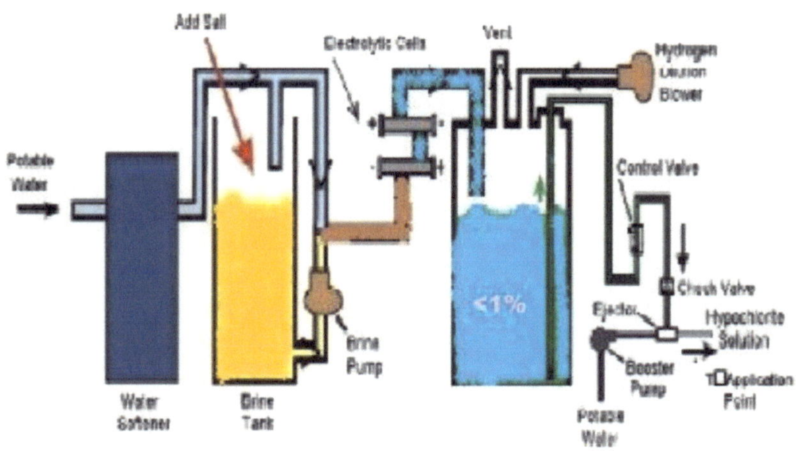

Figure 3. Chlorination by calcium hypochloride method.

2.4.1. *Advantages*

- Chloramine is more stable but not a strong disinfectant as chlorine, providing long lasting residual disinfectant.
- No by-products are formed in chloramination.
- However, the environmental protection agency (EPA) is eager to know about the type and quantity of disinfection byproducts produced by the interactions of chloramines, bromide, brominated organics, and by the chloramination of ozonized waters. EPA's findings may influence chloramine's future use [7].

2.4.2. *Chloramination equipment*

Equipment for both the production of chloramines and chlorination systems are same. Both chlorine and ammonia might be introduced as a liquid or gas. Additionally, both chlorine and ammonia are available in liquid or granular form. Great care must be taken so that concentrated chlorine and ammonia are never mixed as they will form nitrogen trichloride, a potentially explosive compound [11].

2.5. Ozonation

Ozone is an allotropic (unstable) formula of oxygen in which three molecules are combined to produce a new molecule. It quickly decomposes to generate highly reactive free radicals. The ozone's oxidation potential (−2.7 V) is greater than that of the chlorine (−1.36 V) or hypochlorite ion (−1.49 V), substances widely used in wastewater treatment such as oxidants. Ozone is surpassed only by the hydroxyl radical (•OH) and fluoride in its oxidation capacity [7, 12].

2.5.1. *Advantages*

Following are the advantages when ozone is used to treat water:

- Ozone possesses strong oxidizing power
- Short reaction time is needed so germs (including viruses) are killed in a few seconds
- No change in color and taste occurs.
- Requires no chemicals
- Oxygen is provided to water after disinfection
- Destroys and removes algae
- Oxidizes iron and manganese
- Reacts with and removes all organic matter [13].

2.5.2. *Limitation*

Onsite generation is necessary as ozone is unstable at atmospheric pressure. It is toxic in high concentrations as it is a greenhouse gas. The three modules of an ozone plant are ozone destructor, ozone contact chamber, and generator. UV light or the corona discharge process is used for ozone generation. In contact chamber, ozone is added to water. The main purpose of the destructor is to limit the amount of ozone to be stripped out into the air. Three main processes affect the release of ozone after introducing ozone into water: decomposition, reaction with water impurities, and stripping into the atmosphere [14].

2.5.3. *Process*

Ozone is produced from a gas containing oxygen (usually ambient air or pure oxygen). The gas is then passed through an electric field. The air is treated to make sure that it is dry and free from dust impurities. Oxygen is converted to ozone in an electrical field. The ozone is then fed to the contact tank so that ozone is dissolved by water to proceed disinfection process [14].

2.5.4. *Equipment*

The system consists of a combination of oxygen supply and high power electric supply systems: a high voltage electric current flow between electrodes and oxygen is discharged between the electrodes. Electrodes are separated with the help of a dielectric gap that contains the discharge chamber for oxygen flow. Electric field causes the breakdown of oxygen molecules and formation of ozone occurs. After generation, the ozone is directed toward the connection chamber, where the treated water is to be disinfected. The ozone produced will dissociate immediately so onsite generation of ozone must be done.

One of the important features is the contact time as well. So, the ballasting and the deballasting modes of very large ballast amounts of water will be much expensive in full-scale applications. One or several ballast water are equipped with ozone injection equipments that act as contact chambers. For achieving long contact time, that ballast water should be pumped into these contact tanks. In order to reduce the sediment formation and fine particles to enter the treatment process, the ozone treatment must undergo preliminary treatment (filter or cyclone).

The ballast treatment ensures that the all the water present in ballast tanks is exposed to ozone at the beginning of the voyage. Contact time plays an important role in the disinfection process. Disinfection is ensured if long contact is not required against the specific microorganism. If long contact time is required against specific microorganisms, treatment should be done during voyage to enable longer contact times.

Organic carbons are associated with sediment formation. These sediments settle down at the bottom of the ballast tanks during voyage. The sediments contain the microbes that are difficult to treat like bacterial colonies or viral clumps. Ozone treatment might not be efficient in the sediment. Ammonia might be produced as a result of biological activity during voyage. Bromines are produced by the reaction of ammonia and disinfectant residual. Bromines are weaker disinfectants, thus the efficacy is reduced [14].

2.5.5. *Chemical*

Ozone is produced by applying high voltage electricity to a gap (tube) containing pure oxygen or filtered dry air (corona discharge method).

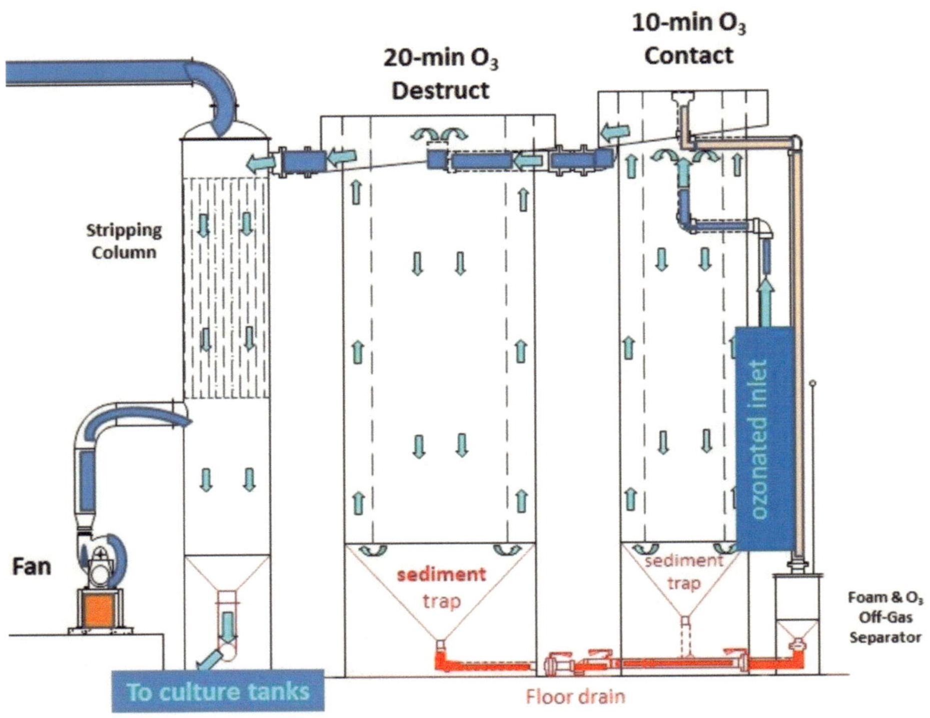

Figure 4. Ozone contacting.

The high voltage electricity results in the formation of ozone by recombining oxygen. The reaction is as follows:

$$3O_2 \rightarrow 2O_3$$

Ozone disinfects by oxidizing the cell walls of microorganisms, which then disintegrates (lyse), destroying the microorganism. This is a very different mechanism than with chlorine, which diffuses through the cell wall, making the cell susceptible to enzymatic attack [7] (**Figure 4**).

2.6. Ultraviolet light (UV)

UV treatment can be used for treating waste water, drinking water, and aquaculture. The UV light causes disinfection by changing the biological components of microorganisms specifically breaking the chemical bonds in DNA, RNA, and proteins [14].

2.6.1. Advantages

- It limits the regrowth potential within the distribution system so no increase in the concentration of biodegradable or assimilable organic carbon (AOC) occurs.

- With respect to interactions with pipe material, there are no concerns.
- No by-products are formed (e.g., hemoglobin-associated acetaldehydes (HAA), trihalomethanes (THM), aldehydes, ketoacidosis, and bromate).
- By using UV light we can achieve the same log inactivation of Giardia and Cryptosporidium, less in cost either than chlorine dioxide and ozone techniques.
- When used in relation with chloramines, no formation of chlorinated disinfection by-product (DBP) is noticed [14].

2.6.2. Limitations

In underdeveloped countries, there are several limitations for UV disinfection. The major limitation is the energy requirement. In many systems, the electric power supply cannot be guaranteed.

A limitation might be that there is not even a single test available to examine the proper disinfection of the rays. It is only effective as a primary disinfectant as it does not leave any residues. It does not act as a secondary disinfectant as it does not work against reinfection in water.

A concern regarding the UV disinfection is chemical composition and the quality of microorganisms present in the influent water. Turbid, cloudy, or the water containing a large number of bacteria can be used to shield bacteria. Chemical composition is a basic problem as the water containing large amount minerals might cause coating on the lamp sleeve, thus reducing the effectiveness of the treatment. Phosphate injectors or water softeners can be used to prevent lamp coating. UV treatment is more effective on low turbid water or partially treated water, which may not be available in the field [15].

2.6.3. Process

UV disinfection units are used nowadays as water disinfection methods. The design is quite simple that consists of a UV light source that is enclosed in a transparent protective sleeve. The light source is mounted so that water can pass through a flow chamber so that UV rays can be both admitted and absorbed into the stream. No change in taste and color occur that is an advantage of this method. The contact time is also very short as these rays kill the pathogenic bacteria quickly [1].

2.6.4. Equipment

UV disinfection systems should be properly shut down if the treatment is not needed for few days. The lamp needs to be warm-up for few minutes before turning on. Moreover, the plumbing system should be properly flushed when not in use. The whole plumbing system should be disinfected by a chemical (preferably chlorine) before relying on the process.

UV lights loses effectiveness with usage, the lamp should, therefore be properly cleaned on regular basis and replaced once in a year. It should be noted that a new lamp might lose its

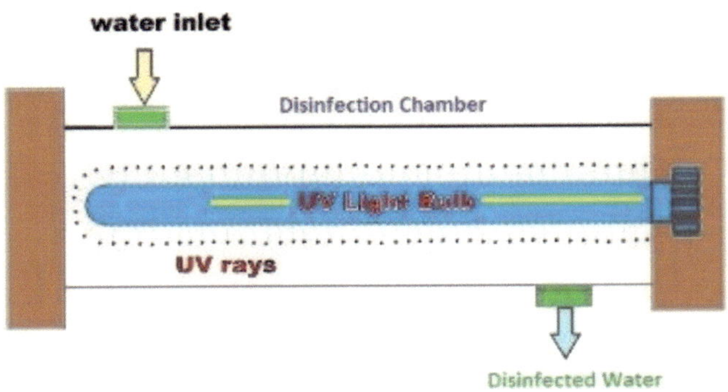

Figure 5. Disinfection by UV.

20% intensity in the first 100 h of operation. Properly calibrated UV detectors help the owner in alerting when the light intensity falls below a certain level.

The UV-treated water must be monitored regularly for the presence of heterotrophic bacteria and coliform bacteria monthly (first 6 months of device's use). The lamp's intensity should be checked if such organisms are noticed [16].

2.6.5. Chemical

UV light can treat water without producing any major chemical or physical changes in the water. No negative effects have been noticed in utilizing UV-treated water. Fewer chances are there for the formation of DBPs as no new substance is added in this process. No change in the taste and color occurs. The dosage and frequency used for the disinfection do not produce any harmful substance. Even the overdosing of UV light does not lead to the formation of harmful products. To avoid exposure, protective clothing should be used by the operator [7] (**Figure 5**).

2.7. Photocatalytic disinfection

The acceleration of a photoreaction in the presence of a catalyst is referred as photocatalysis. In catalyzed photolysis, adsorbed substrate is used to absorb light. In photogenerated catalysis, electron–hole pairs are created by the photocatalytic activity (PCA) generating free radicals (e.g., hydroxyl radicals: •OH) that have the ability to undergo secondary reactions. Its practical application was made possible by the discovering the electrolysis of water by using of titanium dioxide.

2.7.1. Advantages

The followings are the advantages of photocatalytic disinfection

- Photocatalysis uses capacity for renewable and pollution-free solar energy, thus it is a good replacement for the energy-intensive conventional treatment methods.

- In comparison to the conventional treatment methods photocatalysis leads to the formation of harmless compounds.
- Waste water contains different hazardous compounds. Photocatalytic process causes destruction of a wide range of these hazardous compounds in various wastewater streams.
- These reactions are mild. Less chemical input is required and the reaction time is modest.
- It can be applied to hydrogen generation, gaseous phase, and aqueous treatments as well for solid (soil) phase treatments to some extent [17].

2.7.2. Limitations

For the effective TiO_2 application in water treatment, the mass transfer limitation has to be minimized since photocatalytic degradation mainly occurs on the surface of TiO_2. TiO_2 has poor affinity toward organic pollutants (more specifically the hydrophobic organic pollutants) so the adsorption of organic pollutants on the surface of TiO_2 is low that results in slow photocatalytic degradation rates. Therefore, targeting pollutants around the TiO_2 nanoparticles to enhance photocatalytic efficiency require consideration. Besides this, the TiO_2 nanoparticles may undergo aggregation due to the instability of the nanosized particle, which may hamper the light incidence on the active centers and consequently reduction in the catalytic activity occur. However, it should be noted that it may well happen that small particles show higher scattering, which can reduce their photocatalytic activity compared to larger ones. Furthermore, for the slurry system, one main practical challenge to overcome is to recover the nanosized TiO_2 particles from the treated water in regards to both the economic concern and safety concern.

To overcome those limitations of TiO_2-based photocatalysis, the following countermeasures have been adopted in previous studies:

1. Modification of TiO_2 catalyst, in order to achieve the utilization of visible light.
2. The catalyst synthesis should be optimized to obtain catalysts with defined crystal structure, high affinity to various organic pollutants, and smaller particle size.
3. Development and designing of second generation of TiO_2 catalyst, with high separation ability, which can be recovered and regenerated effectively.

The purpose of these modifications and developments is to improve photocatalytic efficiency, complete degradation of organic pollutants, improve visible light absorption, improve stability and reproducibility, and to improve recycle and reuse abilities of TiO_2 [18].

2.7.3. Process

Photocatalytic reaction depends mainly on light (photon) energy or wavelength and the catalyst. Generally, semiconductors are used as catalysts. These materials function as sensitizers for the irradiation of light-stimulated redox process because of their electronic structure. They have a filled valence band and a vacant conduction band.

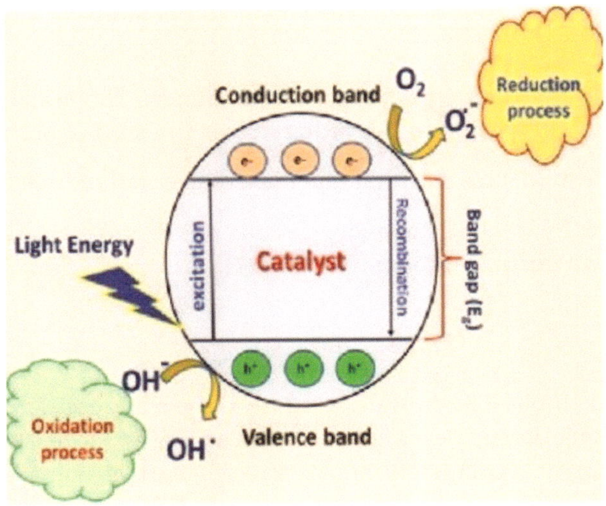

Figure 6. Schematic representation of semiconductor photocatalytic mechanism.

The fundamental steps in the process of semiconductor photocatalysis are as follows:

- When the light energy in terms of photons falls on the surface of a semiconductor and if the energy of incident ray is equivalent or more than the bandgap energy of the semiconductor, the valence band electrons move to the conduction band of the semiconductor.
- The valance band of semiconductors is left with holes. These holes can react with water molecules to generate hydroxyl radicals by oxidizing donor molecules.
- Superoxide ions are formed by reacting the conduction band electrons with dissolved oxygen species. These electrons induce the redox reactions.

These electrons and holes might undergo successive redox reactions with many species to form necessary products by absorbing on the surface of the semiconductor [19] (**Figure 6**).

2.7.4. Chemical

TiO_2 is a semiconductive material that acts as a strong oxidizing agent during illumination by lowering the activation energy required for the decomposition of organic and inorganic compounds. The illumination of the surface of the TiO_2 induces two types of carrier separation: (1) an electron (e−) and (2) a hole (h+). For the production of these two carriers, sufficient amount of energy must be supplied by a photon to move an electron (e−) from the valence band to the conduction band, thus leaving a hole (h+) in the valence band. In comparison to the conducting materials, the recombination of holes and electrons is relatively slow in TiO_2 recombination in metals occurs immediately [20].

$$TiO_2 + hv \rightarrow h^+ + e^-$$

http://dx.doi.org/10.5772/intechopen.80999

3. Conclusions

Water can be affected by environmental factors. Both human and environmental risks are taken into account, which may be tangible and/or intangible. Chlorination can lead to the formation of by-products or toxic chemicals that are hazardous to aquatic life. High chlorine residues may range from avoidance to death of aquatic organisms. The threshold tolerance limit of some aquatic species to chlorine is 0.002 mg/l in freshwater and 0.01 mg/l in saline water. The by-products can also accumulate in the aquatic environment. The toxicity of the chlorinated residues can be eliminated by dechlorination.

In summary, the beneficial use of aquatic ecosystem protection may be compromised when chlorinated wastewater is discharged to receiving surface waters.

Chlorination might not be a risk to the environment if the treated wastewater is reused beneficially rather than discharging into receiving surface waters. An acceptable method for disinfecting wastewater reuse is chlorination. Chlorination is the best method for reuse applications when a residual is residual is required for microbial re-growth. However, there is a limitation of 1 mg/l of chlorine at the point of application of reclaimed water. These limits mostly do not harm the plant life. However, some sensitive crops may be damaged at a level of chlorine lower than 1 mg/l and users should consider the sensitivity of any crops that may be irrigated with chlorine disinfected reclaimed water. However, little environmental risks are associated with the direct use of chlorine. However, the manufacture, storage, and transportation of chlorine products still pose a risk to the environment.

Toxic by-products are formed by the oxidation of ozone. Ozone gas might harm the environment because of its corrosive nature.

Microfiltration only poses a risk to the environment if there is a spill of cleaning agents or the contaminated backwash waste is disposed of incorrectly. UV light poses less risk as compared to other disinfection methods, but it may pose a risk regarding photo-reactivation and mutation of the microbial population present in the discharge. No reuse option is available for UV lamps. Controlling the natural systems like detention lagoons is difficult.

A major environmental risk associated with lagoon-based disinfection is the excessive growth of undesirable organisms, such as blue-green algae. Humans are at high risk as blue-green algal blooms produces toxins. Environment is also at risk as the levels of SS and BOD increases. In terms of potential environmental cost, it would appear that UV, lagoons, and microfiltration have the least potential to impact adversely upon the environment, followed by ozonation and then chlorination. This ranking is based on the formation of by-products and the level of toxicity of the discharge to the receiving environment.

Acknowledgements

The authors thank Ms. Aqdas Zoreen & Jamshaid Khan (Department of Microbiology & Biotechnology) for their sincere help in writing the chapter. The authors also acknowledge the guidance of other faculty members of the department for their guidance and suggestions.

Author details

Muhammad Saqib Ishaq[1*], Zobia Afsheen[2], Amjad Khan[1] and Amjad Khan[3]

*Address all correspondence to: Saqib.ishaq@abasyn.edu.pk

1 Department of Microbiology & Biotechnology, Abasyn University, Peshawar, Khyber Pakhtunkhwa, Pakistan

2 Head of Academics, Akhuwat Faisalabad Institute of Research Science & Technology, Faisalabad, Pakistan

3 Department of Pharmacy, Kohat University of Science & Technology, Kohat, Pakistan

References

[1] Khayyat AMA. Study of point of use treatment methods for the disinfection of drinking water in Nepal [Doctoral dissertation]. Massachusetts Institute of Technology; 2000

[2] U.S. EPA. Wastewater Technology Fact Sheet. EPA 832-F-03-024; 2003

[3] Wagenet L, Darling S, Lemley A. Chlorination of Drinking Water. Fact Sheet 5, September 1988 (Updated October 2005)

[4] Card H, Goulding R, Blanchard C, Penney B, Rumbolt G, Bradley A. A Guide for the Selection of Appropriate Disinfection Equipment for Municipal Applications: Chlorination Equipment Selection Guidelines. New Foundland & Labrador

[5] Lindsay L. Chlorination. Published by The National Environmental Services Center at West Virginia University, Morgantown. 2011;4(2):1-4. Article: Tech Brief • Chlorination, Fall 2004, Vol. 4, Issue 2

[6] Rabêlo Braitt G, de Almeida Rodrigues E, da Silveira Bueno CE, Braitt AH. Evaluation of active chlorine releasing of sodium hypochlorite during seven days, stored at different temperatures. RSBO Revista Sul. 2013. ISSN: Electronic version: 1984-5685, RSBO. 2013 Apr-Jun;10(2):143-8, 7 Pages

[7] Snowden-Swan L, Piatt J, Lesperance A. Disinfection Technologies for Potable Water and Wastewater Treatment: Alternatives to Chlorine Gas. Pacific Northwest National Laboratory (PNNL). Prepared for: US Army Forces Command Air Quality Division, United State;1998

[8] Environmental Protection Agency. Water Treatment Manual Disinfection. Wexford, Ireland: Johnstown Castle, Co.; 2011

[9] Momba MNB, Thompson P, Obi CL. Guidelines for the Improved Disinfection of Small Water Treatment Plants. South Africa: University of South Africa, Water Research Commission. 2008

[10] Tech E. Chlorine and alternative disinfectants guidance manual. Water stewardship-Office of Drinking water. In: Snowden-Swan, L, editor. John Piatt & Ann Lesperance. U.S.: Pacific Northwest National Laboratory; 2005

[11] Arasmith, S. Introduction to small water systems. University of Alaska SE (UAS) Sitka Campus & ATTAC (Alska Training/Technical Assistance Center). United State: ACR Publications. 1993

[12] Forero JE, Duque JJ, Rios F, Diaz J. Ozone for phenol treatment in industrial wastewater. CT&F-Ciencia, Tecnología y Futuro. 2001;2(2):17-26

[13] Brief T. A national drinking water clearinghouse fact sheet by, Mohamed Lahlou. Tech Brief Twelve. 1999

[14] Sassi J, Viitasalo S, Rytkonen J, Leppakoski E. Experiments with ultraviolet light, ultrasound and ozone technologies for onboard ballast water treatment. VTT TIEDOTTEITA. Turku, Finland: Åbo Akademi University; 2005

[15] Mohamad A, Khayyat A. Study of point of use treatment methods for the disinfection of drinking water in Nepal [Thesis]. Brown University, B.Sc. Mechanical Engineering; 1999

[16] Wagenet L, Lemley A. Ultraviolet radiation for disinfecting household drinking water. Cornell Cooperative Extension, New York State College of Human Ecology, Cornell University, U.S; 1993. ISSN: Electronic version: 1984-5685, RSBO. 2013 Apr-Jun;**10**(2):143-8, 7 Pages

[17] Saravanan R, Gracia F, Stephen A. Basic principles, mechanism, and challenges of photocatalysis. In: Nanocomposites for Visible Light-induced Photocatalysis. Cham: Springer; 2017. pp. 19-40

[18] Dong H, Zeng G, Tang L, Fan C, Zhang C, He X, et al. An overview on limitations of TiO_2-based particles for photocatalytic degradation of organic pollutants and the corresponding countermeasures. Water Research. 2015;**79**:128-146

[19] Lazar MA, Varghese S, Nair SS. Photocatalytic water treatment by titanium dioxide: Recent updates. Catalysts. 2012;**2**(4):572-601

[20] Sokolowski A. Effects of nanostructured TiO_2 photocatalysis on disinfection by-product formation [Doctoral dissertation]. Canada: University of Toronto; 2014

Chapter 2

Selective Photodegradation Using Titanate Nanostructures

Ayman Hassan Zaki and
Waleed Mohamed Ali. El Rouby

Additional information is available at the end of the chapter

http://dx.doi.org/10.5772/intechopen.80311

Abstract

Adsorption and photocatalytic degradation are considered as the most important ways of treating water from organic compounds. It would be very useful if the adsorption and photocatalytic properties are combined in the same catalyst used in the treatment. Titania is one of the best well-known photocatalysts. However, due to its poor selectivity, it is unfavorable for photocatalytic removal of highly toxic low-level organic pollutants in wastewater in the presence of other less toxic high-level pollutants. Recent trials to introduce selectivity for titania have been achieved via controlling the catalyst morphology or by modifying the catalyst surface. This chapter summarizes the control of selectivity of titanate nanostructures toward adsorption and/or photocatalytic degradation of toxic organic dyes. In the first part, the effect of morphologies of titanites on selective photocatalytic degradation of three food dyes (color yellow sunset, red allura, and red carmoisine) was discussed. In changing the morphology of titanite, each dye is being preferably adsorbed by one morphology and decomposing more rapidly. In the second part, the selective adsorption and/or photocatalytic degradation of methylene blue dye from mixed dye solution using sodium titanate (NaTNT), cobalt-doped titanate nanotubes (co-doped TNT), and the decorated one with gold nanoparticles has been discussed.

Keywords: titanate, selectivity, TiO_2, photocatalysis, dyes, water treatment

1. Introduction

Advanced oxidation process (AOP) is considered as one of the innovative technologies that is used for degradation of various organic pollutants by generating highly reactive species [1]. Heterogeneous photocatalysis is one of the most effective methods leading to complete mineralization of organic pollutants to carbon dioxide, water, or low molecular weight compounds under visible or ultraviolet (UV) light radiation [2–5]. Photocatalytic degradation of various families of organic pollutants had been studied using semiconductors such as TiO_2, ZnO, Fe_2O_3, CdS, GaP, and ZnS [6–15]. Titanium dioxide (TiO_2) is considered as one of the most important photocatalysts used in water treatment application where it is stable, inexpensive, nontoxic, insoluble, and potentially reusable in water [2, 16–19].

As it is well known, the mechanism of degradation of an organic pollutant using TiO_2 nanostructures depends mainly on producing hydroxyl radicals ($^{\cdot}OH$). TiO_2 absorbs in the UV region and produces electron-hole pairs. The electron-hole (e^--h^+) pairs generate superoxide ions ($O2^{\cdot -}$), peroxide radicals ($^{\cdot}OOH$), hydroxyl ions (OH^-), and H^+ ions. All these species will finally form the desired hydroxyl radicals ($^{\cdot}OH$). Free radicals are aggressive species and highly active in the chemical reactions; these radicals attack the organic pollutants forming different oxygenated intermediates and finally converting them into CO_2 and H_2O.

TiO_2 with variable morphologies, rods, spheres, tubes, fibers, sheets, and interconnected architectures, can be prepared with many methods such as hydrothermal method, sol-gel method, electrodeposition, chemical vapor deposition, and microwave method [20, 21].

The photocatalytic activity of TiO_2 depends on many factors such as specific surface area, size, pore structure, pore volume, exposed surface facet, and crystalline phase [20, 22, 23]; achieving selectivity in degradation using TiO_2 in some cases depends on crystallinity and crystal facet [24–26], and also depends on introducing molecular sites on TiO_2 surface which produce selective adsorption and degradation for targeted compounds are also introduced [27].

2. Selective photocatalytic degradation of dyes

2.1. Selective photocatalytic degradation by controlling titanate morphology and crystal structure

The adsorption stage is a very important stage in the photocatalytic process; many factors affect on the selectivity in this stage like changing the size, amount, surface, and morphology of the photocatalyst, as well as the size or type of the target compounds. Recent studies revealed that altering the exposed surfaces of TiO_2 nanoparticles is by controlling the morphologies of these particles. Sofianou et al. [25] found that the calcined TiO_2 nanoplates showed the highest photocatalytic activity toward oxidizing NO gas to NO_2 and NO_3^-; on the other hand, the washed TiO_2 nanoplates, preserving the initial morphology, showed the best photocatalytic activity toward acetaldehyde decomposition. It was concluded that the dominant exposed {1 0 1} or {0 0 1} crystal facet of the TiO_2 nanoplates is considered as the key factor in controlling the selectivity in adsorption of these air pollutants. Also, it was demonstrated by other researchers that modifying the surface of TiO_2 microspheres by varying the degree of the etching of {0 0 1} facets exhibited tunable photocatalytic selectivity toward the decomposition of azo dyes in water [28].

Recently, Zaki et al. [22] used three different colors (color yellow sunset, color red allura, and color red carmoisine) to test the selectivity of degradation of three different morphologies of TiO_2 (spherical, layered, and tubular); the TEM images of these morphologies are shown in **Figure 1**. It is clear from **Figure 1a** that the starting TiO_2 powder consists of nanosized particles, while **Figure 1b** shows the TEM image of the synthesized TiO_2 nanosheets, and **Figure 1c** shows the TEM images of the obtained TiO_2 nanotubes. The tubes have a diameter of about 16 nm, and they are randomly oriented with nearly homogenous dimensions with some intercalated tubes.

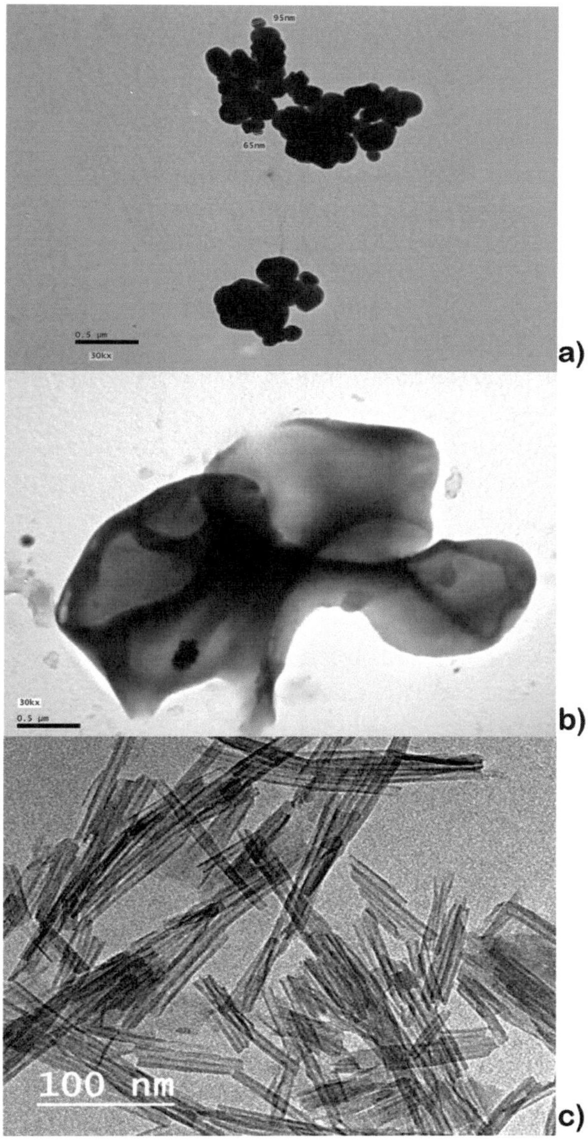

Figure 1. TEM image of spherical TiO_2 (a), TiO_2 nanosheets (b), and TiO_2 nanotubes (c).

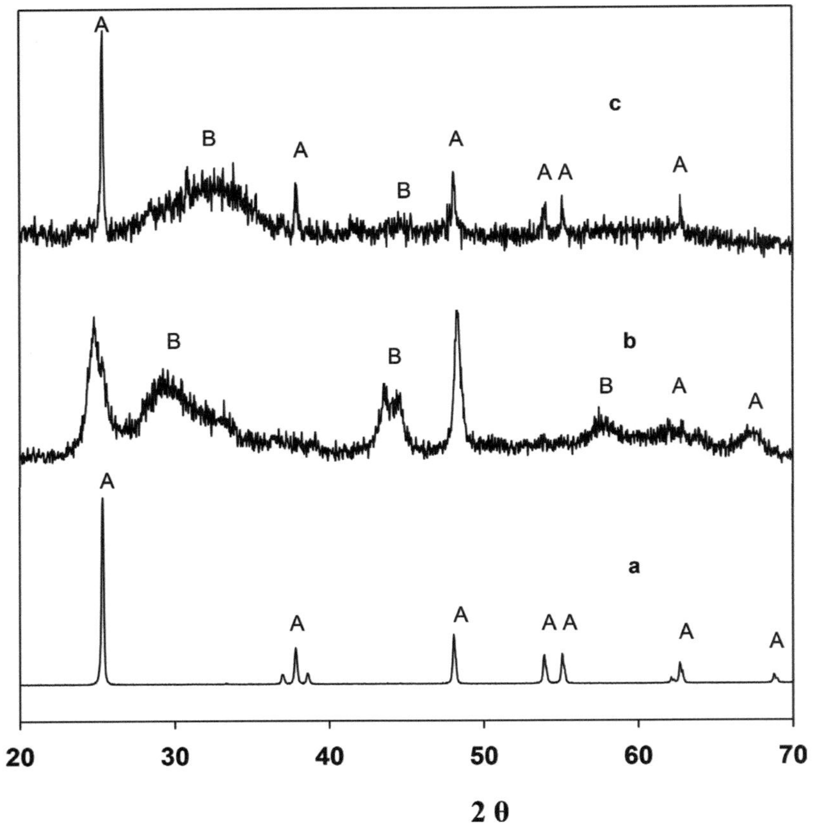

Figure 2. XRD patterns of spherical TiO_2 (a), TiO_2 nanotubes (b), and TiO_2 nanosheets (c). A, anatase; B, TiO_2 (B).

Figure 2 shows XRD patterns of the three morphologies; it is clear from these patterns that the crystallinity of TiO_2 nanotubes and TiO_2 nanosheets is generally poor; this may be attributed to the small sizes of the prepared samples, which are confirmed by the presence of broad peaks. All the detected peaks confirmed the presence of anatase phase for spherical TiO_2 nanoparticles; in the case of nanosheets and nanotubes, there is a contribution of TiO_2 (B) with different ratios depending on the morphology; accordingly, we have preferred three. The crystal sizes of the three morphologies were calculated using Scherrer's formula and are listed in **Table 1**.

Catalyst		*K* (min^{-1})		
Morphology	**Crystal size (nm)**	*K* (min^{-1}) *for* **Yellow sunset**	*K* (min^{-1}) *for* **Red allura**	*K* (min^{-1}) *for* **Red carmoisine**
Spherical TiO_2	95 nm	0.0065	0.0098	0.0082
TiO_2 nanosheets	67.9 nm	0.0513	0.0059	0.0103
TiO_2 nanotubes	27.1 nm	0.0324	0.0060	0.0189

Table 1. Structural and kinetic parameter TiO_2 nanostructures.

Zaki et al. in this work reported that changing the morphology of TiO_2 from spherical to layered and tubular shape made each morphology preferentially decompose one dye of the three dyes [22]. From **Figure 3a–c**, it is clear that the photocatalytic degradation strongly depends on the morphology of TiO_2. To test photocatalytic performance of all morphologies, they started the test with yellow sunset as a model, and they found that as moving from the spherical to the tubular structure via the sheet structure, the time of degradation reduced from 400 min for spherical, 75 min for nanotubes, and 55 min for nanosheets, while in the case of the other two colors, red allura and red carmoisine, they found other trends, as shown in **Figure 4a–c**, where the best

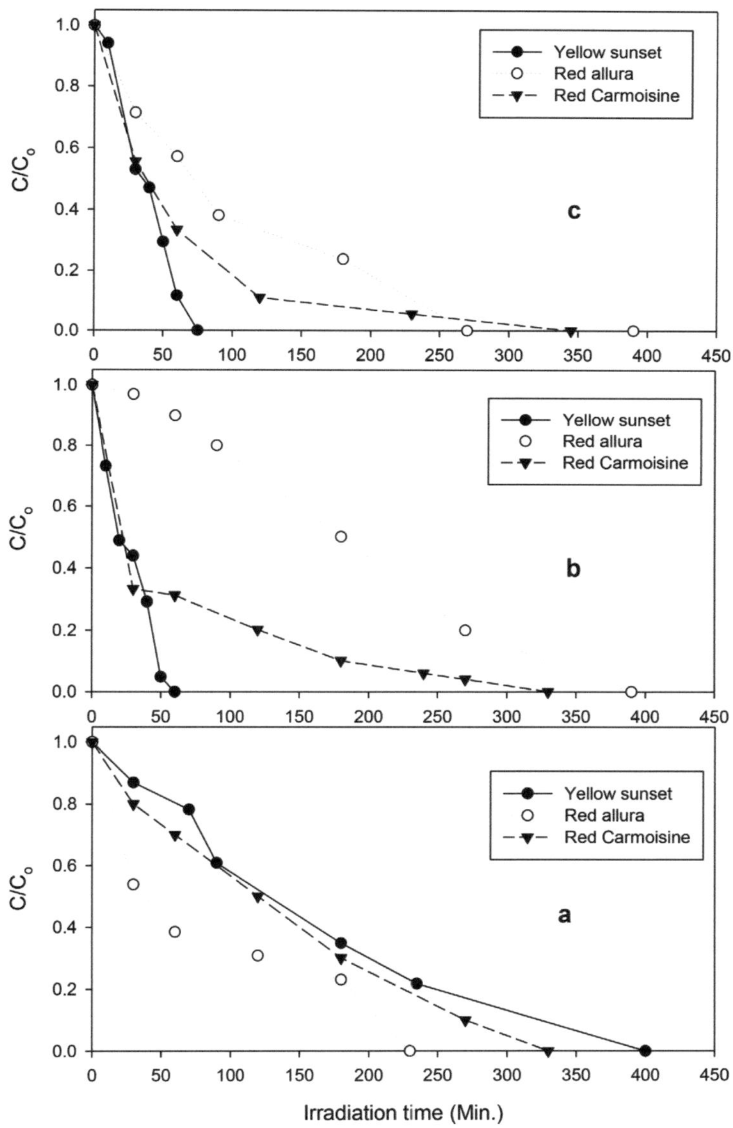

Figure 3. Photocatalytic activity of spherical (a) TiO_2 nanosheets and (c) TiO_2 nanotubes.

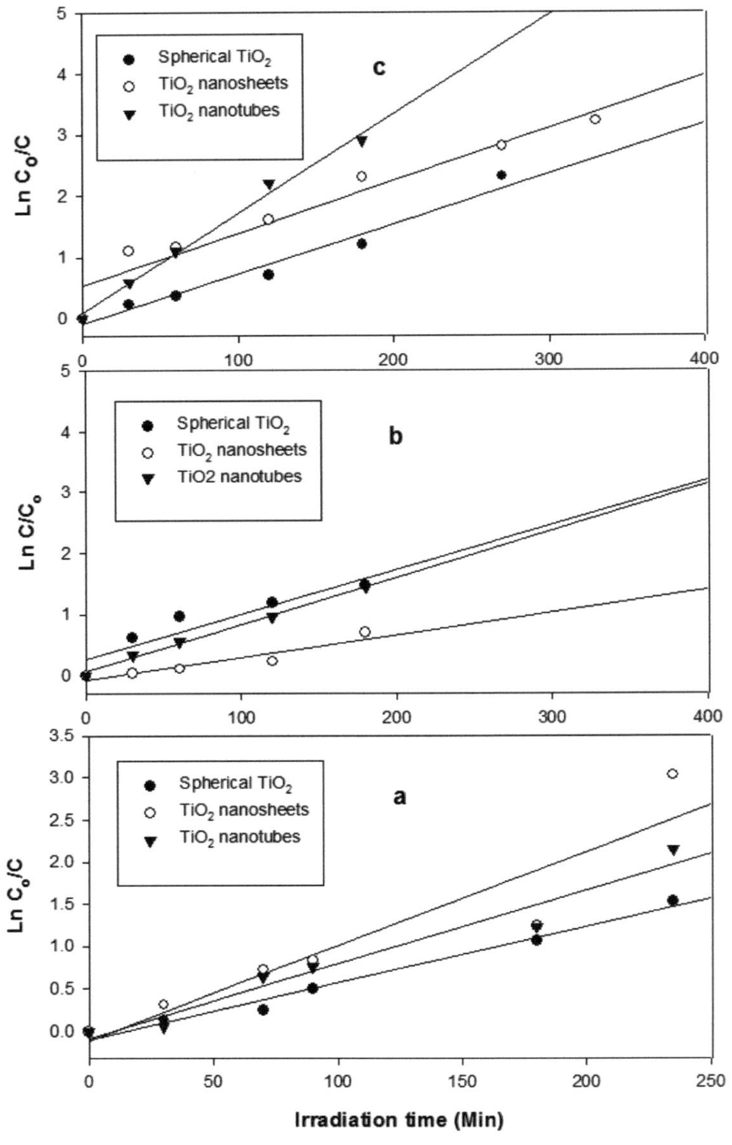

Figure 4. Linear transform ln (Co/C) = f(t) of kinetic curves.

degradation rate for yellow sunset was achieved by TiO_2 nanosheets, by spherical TiO_2 for allura and by TiO_2 nanotubes for carmoisine. Finally, they concluded that the preferred orientation of each morphology made it more specific in action; each dye is adsorbed preferentially by one of the three morphologies and decomposed more rapidly.

3. Selective adsorption and degradation over decorated titanate nanotubes

There is a major concern on contamination of water by organic pollutants [29–31]. Nowadays, intensive efforts have been devoted for water treatment especially when the problem of water shortage has begun to loom in some cities and countries, particularly in Africa and the Middle East. This problem has made scientists think of reusing water again after treatment. Therefore, many methods of treatment have been used such as filtration [32], biodegradation [33], adsorption [34], and photocatalytic degradation [35] of contaminant especially the organic molecules. Among these methods, adsorption is the simplest, cheapest, and most versatile technique for treating water pollutants [30, 36]. Photocatalytic degradation is also considered as one of the promising technologies used for wastewater treatment, where a suitable catalyst is used for the degradation of toxic organic molecules under irradiation with light [35, 37, 38].

Titanium dioxide (TiO_2) is considered as one of the most important photocatalysts used in water treatment application where it is stable, inexpensive, nontoxic, and potentially reusable in water; however, its adsorption ability is very low. Therefore, the key challenge of using TiO_2 for the treatment of industrial dye-mediated wastewater is to provide adsorption characteristics through surface or structure modification. Another drawback of TiO_2 is the low selectivity due to the formed reactive species (radicals) by means of light irradiation which is difficult to be controlled. In addition, TiO_2 efficiently degrades organic pollutants but requires ultraviolet light for activation. Thus, the ideal solution to it is to acquire TiO_2 selective ***adsorption*** property which will be a powerful technique for imparting selectivity to photocatalytic degradation also.

Transformation of TiO_2 nanoparticles to the tubular titanate forms could give some unique photocatalytic properties because of the one-dimensional (1D) geometry, which will enable the electron transfer faster for long distance. Moreover, its nanotubular structure will provide a large specific surface area and pore volume which is very important for providing more active sites for adsorption and photocatalytic degradation.

The treatment of highly enduring toxic dyes and leaving the alterable pollutants to be treated by the low-cost biological treatment systems are considered as great aims for scientists interested in wastewater treatment. Thus, provision of selective materials may be of great benefit. Therefore, one of the important challenges in water treatment is to possess a highly selective method for adsorption and photodegradation of contaminants [39].

Titanate nanotubes (TNTs) produced from the hydrothermal treatment of TiO_2 nanoparticles in the presence of high concentration of NaOH (10 M) at 160°C have been used for selective adsorption of specific dyes from water [40, 41]. The obtained sodium titanate (Na_2TiO_3, NaTNT) nanotubes were characterized before using as an adsorbent. In the TEM micrograph (**Figure 5a** and **b**), a clear tubular structure of about 5 nm inner cavity with a mean length of 148 ($\pm$35) nm and thickness of 8 ($\pm$1) nm is seen. Due to NaTNTs having a large bandgap (more than 3 ev), it needs high-energy light (UV light) for initiating its photocatalytic activity. Thus, for improving the photocatalytic properties of NaTNTs, gold (Au) nanoparticles were deposited on the surface where the Au nanoparticles have a plasmonic response in the visible light. The in situ photoreduction deposition method was used for forming a good contact between the gold nanoparticles and NaTNTs. From **Figure 5c** and **d**, a homogeneous distribution of highly crystalline spherical Au nanoparticles with a mean diameter of 7.7 (±1) nm was formed at the surface of the NaTNTs, and there is no any aggregation of Au particles.

As mentioned before, the adsorption ability of TiO_2 nanoparticles is extremely low. El Rouby et al. reported that the titanate form can adsorb up to 55% of the methylene blue (MB) dye initial concentration after 180 min compared to the TiO_2 nanoparticles which show negligible adsorption for MB (**Figure 6a**) [40]. This high adsorption ability of NaTNTs is due to their high porosity, large specific surface area, and strong electrostatic interaction between the positively charged dye and the negatively charged surface of NaTNTs. In addition, the internal cavity of the tubular structure can contribute to further extend the area on which organic molecules can be adsorbed.

The photocatalytic activity of NaTNTs and Au-decorated NaTNTs has been assessed under simulated solar light (350–2400 nm). Twenty-seven percent degradation of MB was achieved

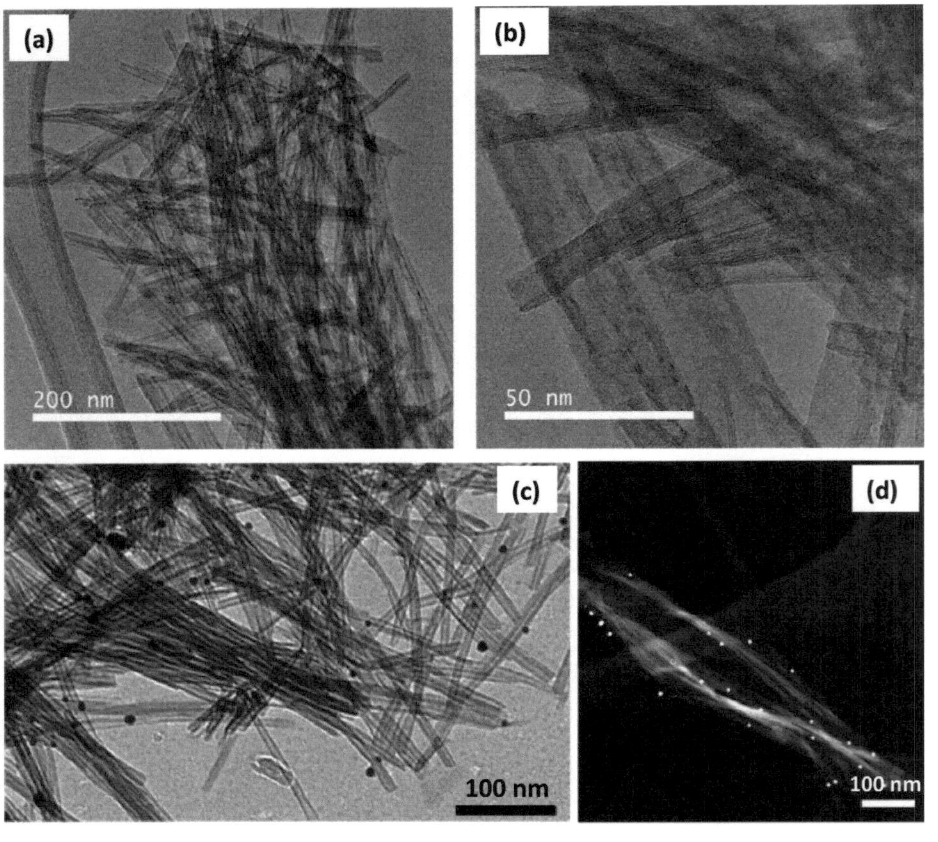

Figure 5. (a) Low and (b) high magnification TEM images of the NaTNTs, (c) TEM images of au-NaTNTs, and (d) STEM images of the Au-NaTNTs [12] (copyright 2017, IOP Publishing Ltd).

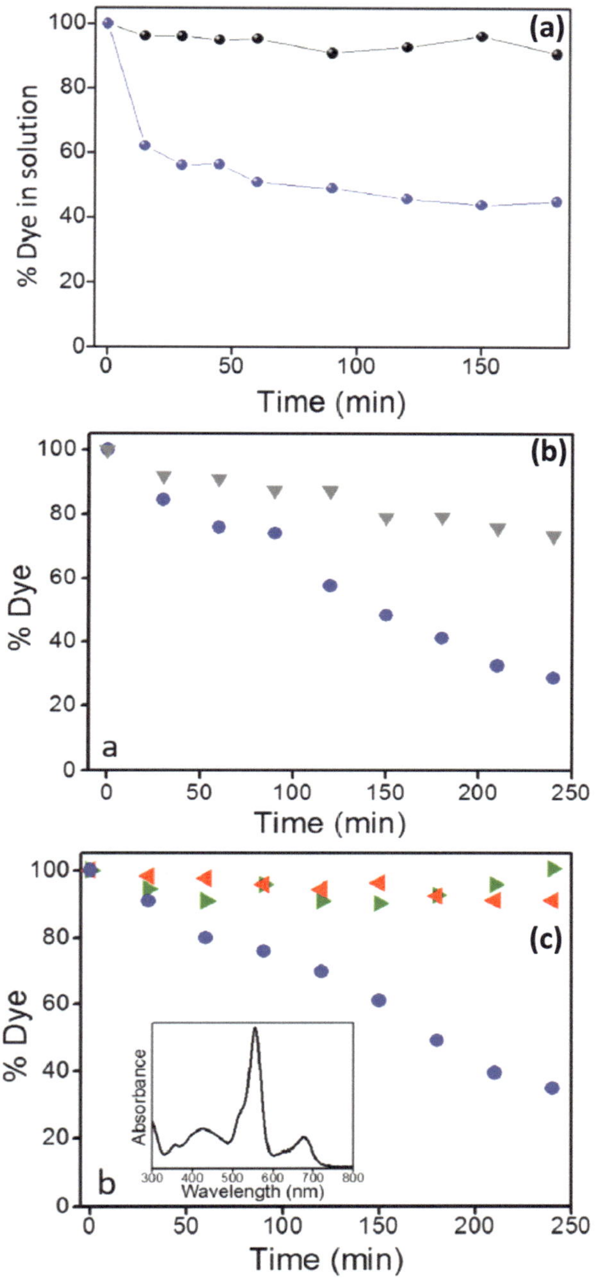

Figure 6. (a) Effect of contact time on MB dye adsorption in the presence of TiO_2 nanoparticles (black) and NaTNTs (blue) under agitation in the dark. (b) Photodegradation of MB in the presence of NaTNTs (gray triangles) and Au-functionalized NaTNTs (blue circles). (c) Photodegradation of the three mixed organic dyes: Tz (green triangles), RhB (red triangles), and MB (blue circles) using Au-functionalized NaTNTs as catalysts. The inset shows the absorption spectrum of the mixture after adsorption/desorption equilibrium and prior to the photocatalytic degradation. The absorption maxima for each dye in such a solution are 421 nm (Tz), 555 nm (RhB), and 674 nm (MB) [12] (copyright 2017, IOP Publishing Ltd).

by NaTNTs that may be ascribed to the small portion of UV light irradiated from the solar simulator (350–400 nm). When Au-decorated NaTNTs were used as a photocatalyst, a remarkable 72% degradation of MB is observed (**Figure 6b**). This is due to the plasmonic photosensitization of gold nanoparticle which is considered as an effective way for increasing the photocatalytic activity of semiconductors holding large bandgaps such as TNTs.

The well-established Au-decorated NaTNT photocatalyst was applied for selective degradation of MB from dye mixture. Three different dyes as model (MB, rhodamine B (RhB), and tartrazine (Tz)) have been used where the absorption spectra of the three dyes allow for interference-free monitoring of the photodegradation for each one of them, thus facilitating their tracking over time through the change in its absorbance. As shown in **Figure 6c**, 65% of MB was decomposed after 240 min, while Tz and RhB remain in solution showing no trace of reaction after that time. These results clearly evidence for the selectivity of Au-NaTNT composite to MB degradation. This selectivity toward MB is attributed to the net charge of the dyes under consideration [40]. Through electrotactic interaction, the surface of the NaTNTs governs the preferential adsorption of the positively charged dye (MB), thus facilitating a high degree of contact between MB and the photocatalyst. On the other hand, in the case of Tz, it remains intact in solution due to its negatively charged surface which makes repulsion with the negative charges on NaTNTs. In a similar way, the interaction of RhB with the NaTNT surface is significantly weakened, due to the deprotonation of the RhB carboxyl group when dissolved in water and thus the formation of a zwitterionic species [40].

For the photocatalytic degradation of organic dyes in a highly concentrated solution, it will be difficult for the catalyst to achieve degradation unless it has a good adsorption property. This is because during the photocatalytic process by any semiconductor, the light strikes the catalyst surface leading to excitation of valence band electrons forming holes in the valence band that can make degradation for the pollutants. But in the case of high-concentrated dye solution, the dyes can strongly absorb the incident light and prevent the light to reach the catalyst surface forming impermeable solution. Thus, the electrons cannot be exited, and the degradation process will not start. Thus, in that case, it is very useful to use a catalyst with high adsorption property especially in addition to its high photocatalytic degradation properties.

In another work, El Rouby [41] tried to increase the adsorption ability of NaTNTs through doping by transition metal (Co). It was found that replacement of Na^+ by Co^{2+} leads to increase the specific surface area as shown in **Table 2** which will directly affect the adsorption extent.

Sample	BET (m^2/g)	Surface area of pores (m^2/g)	Pore size (Å)	Pore volume (cm^3/g)
NaTNT	132.15	144.37	51.09	0.900
Co-doped TNT 1	161.47	163.25	53.23	0.243
Co-doped TNT 2	162.62	177.94	49.33	0.226
Co-doped TNT 3	170.83	185.14	51.62	0.255

Table 2. Effect contents on surface area, pore size, and pore volume of Co-doped TNT samples [13] (copyright 2018, Elsevier).

This increase of specific surface area is due to the formation of a large number of pores because of the removal of sodium ions (larger ions) and insertion of cobalt ions (smaller ions). This leads to the increase in the surface area of pore as the cobalt content was increased in the samples (**Table 2**). It was found that MB uptake on NaTNTs was about 89 mg/g after 35 min. This indicated its ferocity to remove MB from water in a very short time at room temperature. In replacing the Na^+ by Co^{2+}, the surface area and number of pores were increased as mentioned previously. Thus, the available adsorption active sites will be more than in the case of NaTNT. This was reflected on the adsorption capacity of Co-doped TNT samples, where the uptake was increased by increasing the dopant content. The adsorption capacity reached 92.5, 91.9, and 91.8 for Co-doped TNT 1, Co-doped TNT 2, and Co-doped TNT 3, respectively, depending on the concentration of Co^{2+} in NaTNTs (**Figure 7**).

The problem in Co-doped TNT catalysts is its wide bandgap as in the cases of TiO_2 and NaTNTs. So, for adding a photocatalytic property, the prepared Co-doped TNTs were calcined at 500°C for enhancing its crystallinity and thus the photocatalytic activity. But, after calcination, the adsorption property was decreased, and the degradation was low (**Figure 8a**) due to light shielding by the high concentration of MB. The decrease in adsorption is due to the decrease in surface area which is caused by means of wall collapse and loose of tubular structure after calcination. Therefore, the previous concept of attaching Au nanoparticles to the surface of Co-doped TNTs was used. It was noticed that MB concentration (1×10^{-4} M) was decreased gradually with time up to (1.2×10^{-5} M) after 300 min, and with continuous illumination, MB was completely degraded (**Figure 8b**). The conclusion from these results is that a highly concentrated organic dye and pollutant solution can be photocatalytically degraded efficiently [41].

The prepared Au-decorated Co-doped TNT catalysts were used for the selective adsorption and degradation of MB in the same dye mixture (MB, RhB, TZ). From **Figure 9a**, the selective

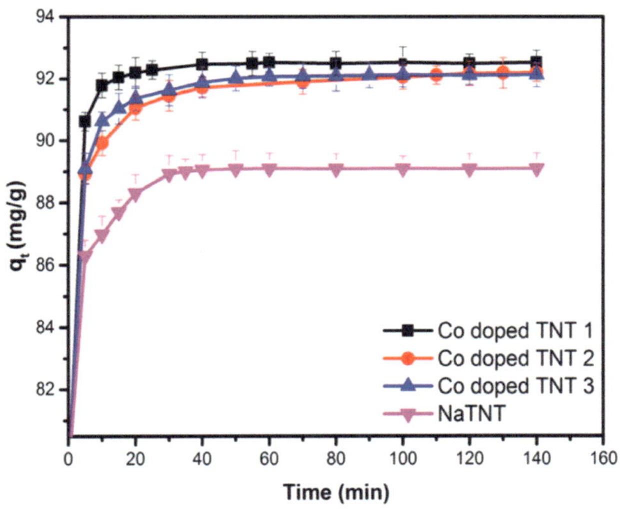

Figure 7. Methylene blue dye (1×10^{-4} M) uptake on titanate nanotubes doped with different ratios of cobalt (0.4 g/L) [13] (copyright 2018, Elsevier).

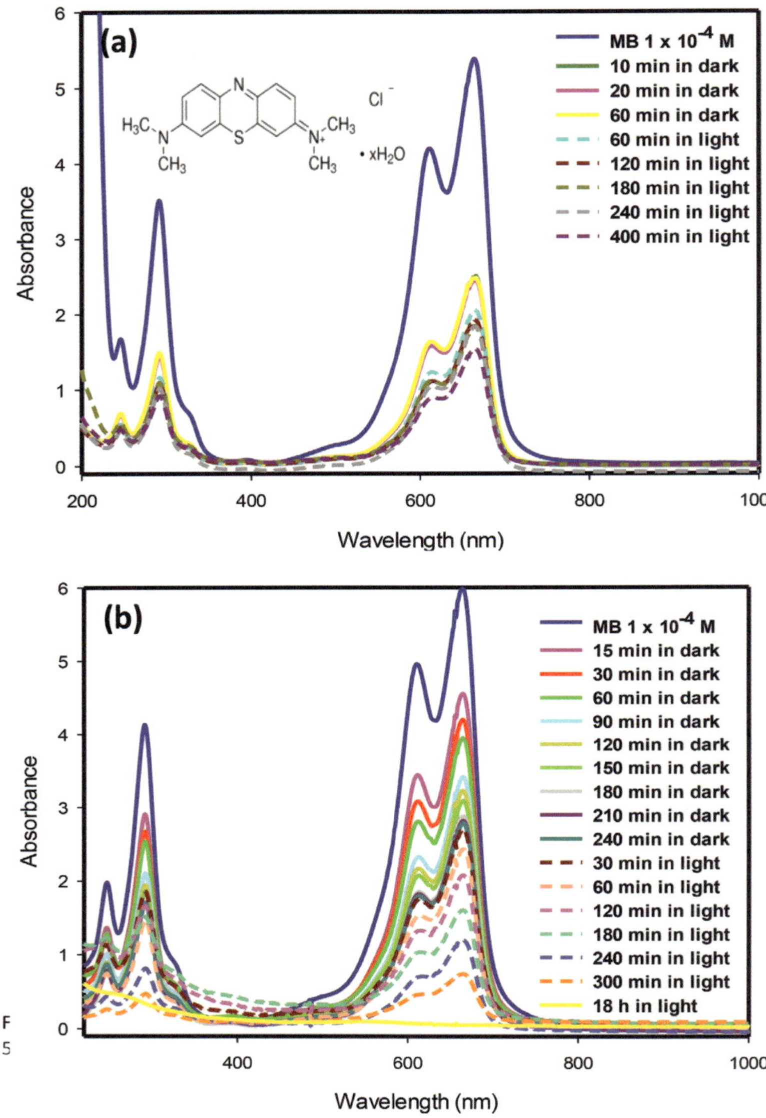

Figure 8. Absorption spectra of MB dye (1×10^{-4} M) in presence of (a) Co-doped TNT 1 calcined at 500°C for 2 h and (b) Co-doped TNT 1 at Au in dark and under illumination [13] (copyright 2018, Elsevier).

adsorption of MB from the dye mixture is clear for the same reason of surface charge as discussed before. After adsorption and subjecting the dye solutions to illumination of simulated light, the degradation of MB was started (**Figure 9b**). According to postulates of any photocatalytic process, the catalyst will initially adsorb the reactant, catalyze the reaction, and finally leave the products. By the same concept, the adsorbed dye was degraded, while dyes in the solution remained without effect [13].

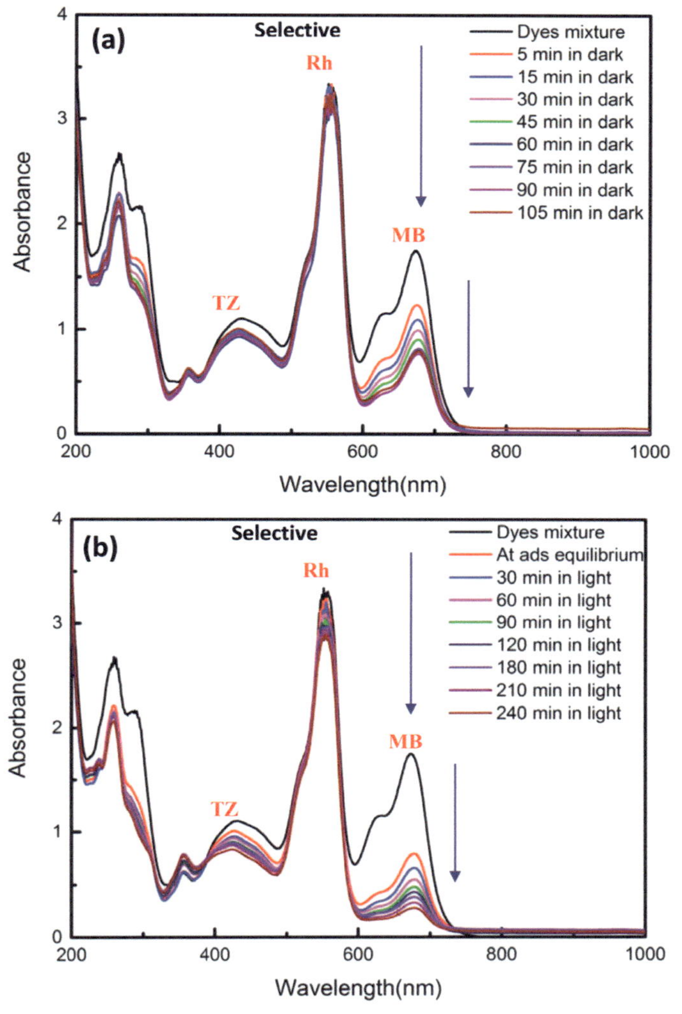

Figure 9. Selective adsorption (a) and selective degradation (b) of MB dye from dye mixture (methylene blue, rhodamine B, and tartrazine dye [5×10^{-5} M]) in dark and under simulated solar light illumination using Au-decorated Co-doped TNT 1 as a catalyst (0.4 g/L) [13] (copyright 2018, Elsevier).

Author details

Ayman Hassan Zaki* and Waleed Mohamed Ali. El Rouby

*Address all correspondence to: ayman_h_zaki@yahoo.com

Materials Science and Nanotechnology Department, Faculty of Postgraduate Studies for Advanced Sciences (PSAS), Beni-Suef University, Beni-Suef, Egypt

References

[1] Park H, Park Y, Kim W, Choi W. Surface modification of TiO_2 photo-catalyst for environmental applications. Journal of Photochemistry and Photobiology C: Photochemistry Reviews. 2013;**15**:1-20

[2] Gupta AK, Pal A, Sahoo C. Photocatalytic degradation of a mixture of crystal violet (basic violet 3) and methyl red dye in aqueous suspensions using Ag+ doped TiO_2. Dyes and Pigments. 2006;**69**(3):224-232

[3] Chen F, Fang P, Gao Y, Liu Z, Liu Y, Dai Y. Effective removal of high-chroma crystal violet over TiO_2-based nanosheet by adsorption–photocatalytic degradation. Chemical Engineering Journal. 2012;**204**:107-113

[4] Liu FY, Jiang YR, Chen CC, Lee WW. Novel synthesis of PbBiO2Cl/BiOCl nanocomposite with enhanced visible-driven-light photocatalytic activity. Catalysis Today. 2018;**300**:112-123

[5] Attia EF, Zaki AH, El-Dek SI, Farghali AA. Synthesis, physicochemical properties and photocatalytic activity of nanosized Mg doped Mn ferrite. Journal of Molecular Liquids. 2017;**231**:589-596

[6] Chakrabarti S, Dutta BK. Photocatalytic degradation of model textile dyes in wastewater using ZnO as semiconductor catalyst. Journal of Hazardous Materials. 2004;**112**(3):269-278

[7] Reddy MP, Venugopal A, Subrahmanyam M. Hydroxyapatite photocatalytic degradation of calmagite (an azo dye) in aqueous suspension. Applied Catalysis B: Environmental. 2007;**69**(3):164-170

[8] Saquib M, Tariq MA, Faisal M, Muneer M. Photocatalytic degradation of two selected dye derivatives in aqueous suspensions of titanium dioxide. Desalination. 2008;**219**(1–3):301-311

[9] Abdel-Khalek AA, Mahmoud SA, Zaki AH. Visible light assisted photocatalytic degradation of crystal violet, bromophenol blue and eosin Y dyes using AgBr-ZnO nanocomposite. Environmental Nanotechnology, Monitoring & Management. 2018;**9**:164-173

[10] Stylidi M, Kondarides DI, Verykios XE. Visible light-induced photo-catalytic degradation of acid Orange 7 in aqueous TiO_2 suspensions. Applied Catalysis B: Environmental. 2004; **47**(3):189-201

[11] Su C, Hong BY, Tseng CM. Sol–gel preparation and photocatalysis of titanium dioxide. Catalysis Today. 2004;**96**(3):119-126

[12] Sun J, Wang X, Sun J, Sun R, Sun S, Qiao L. Photocatalytic degradation and kinetics of Orange G using nano-sized Sn (IV)/TiO_2/AC photocatalyst. Journal of Molecular Catalysis A: Chemical. 2006;**260**(1):241-246

[13] Herrmann JM. Water treatment by heterogeneous photocatalysis. Catalytic Science Series. 1999;**1**:171-194

[14] Legrini O, Oliveros E, Braun AM. Photochemical processes for water treatment. Chemical Reviews. 1993;**93**(2):671-698

[15] Blake D. Bibliography of Work on the Heterogeneous Photocatalytic Removal of Hazardous Compounds from Water and Air. Update Number 4 to October 2001 (No. NREL/TP-510-31319). Golden, CO (US): National Renewable Energy Laboratory; 2001

[16] Bavykin DV, Friedrich JM, Walsh FC. Protonated titanates and TiO_2 nanostructured materials: Synthesis, properties, and applications. Advanced Materials. 2006;**18**(21):2807-2824

[17] Nakata K, Fujishima A. TiO_2 photocatalysis: Design and applications. Journal of Photochemistry and Photobiology C: Photochemistry Reviews. 2012;**13**(3):169-189

[18] Zaki AH et al. Morphology transformation from titanate nanotubes to TiO2 microspheres. Materials Science in Semiconductor Processing. 2018;**75**:10-17

[19] Shaker AM, El-Shahawy A, Zaki AH, Abdel-Rahim EF, Khedr MH. Estimation the median lethal dose and inhibitory concentration of TiO_2, SiO_2, ZnO and CuO nanoparticles on human hepatoma HEPG2 cells. International Journal of Pharmaceutical and Phytopharmacological Research. 2017;**7**(6):18-23

[20] Chen X, Mao SS. Titanium dioxide nanomaterials: Synthesis, properties, modifications, and applications. Chemical Reviews. 2007;**107**(7):2891-2959

[21] Tayade RJ, Key DL. Synthesis and characterization of titanium dioxide nanotubes for photocatalytic degradation of aqueous nitrobenzene in the presence of sunlight. In: Materials Science Forum. Trans Tech Publications; 2010. Vol. 657. pp. 62-74

[22] Farghali AA, Zaki AH, Khedr MH. Control of selectivity in heterogeneous photocatalysis by tuning TiO2 morphology for water treatment applications. Nanomaterials and Nanotechnology. 2016;**6**:12

[23] Lazar MA, Tayade RJ, Bajaj HC, Jasra RV. Correlation of surface properties and photocatalytic activity of nanocrystalline TiO2 on the synthesis route. In: Nano Hybrids. Trans Tech Publications; 2012. Vol. 1. pp. 57-80

[24] Lazar MA, Daoud WA. Achieving selectivity in TiO2-based photocatalysis. RSC Advances. 2013;**3**(13):4130-4140

[25] Sofianou MV, Psycharis V, Boukos N, Vaimakis T, Yu J, Dillert R, et al. Tuning the photocatalytic selectivity of TiO2 anatase nanoplates by altering the exposed crystal facets content. Applied Catalysis B: Environmental. 2013;**142**:761-768

[26] Zhang J, Chen S, Qian L, Tao X, Yang L, Wang H, et al. Regulating photocatalytic selectivity of anatase TiO2 with {101}, {001}, and {111} facets. Journal of the American Ceramic Society. 2014;**97**(12):4005-4010

[27] Tominaga Y, Kubo T, Hosoya K. Surface modification of TiO2 for selective photodegradation of toxic compounds. Catalysis Communications. 2011;**12**(9):785-789

[28] Nguyen-Phan TD, Shin EW. Morphological effect of TiO_2 catalysts on photocatalytic degradation of methylene blue. Journal of Industrial and Engineering Chemistry. 2011; **17**(3):397-400

[29] Laurier KGM, Vermoortele F, Ameloot R, De Vos DE, Hofkens J, Roeffaers MBJ. Iron (III)-based metal–organic frameworks as visible light photocatalysts. Journal of the American Chemical Society. 2013;**135**:14488-14491

[30] Farghali AA, Bahgat M, El Rouby WMA, Khedr MH. Decoration of MWCNTs with $CoFe_2O_4$ nanoparticles for methylene blue dye adsorption. Journal of Solution Chemistry. 2012;**41**(12):2209-2225

[31] Sillanpää M, Ncibi MC, Matilainen A, Vepsäläinen M. Removal of natural organic matter in drinking water treatment by coagulation: A comprehensive review. Chemosphere. 2018;**190**:54-71

[32] Yoo SS, Chu KH, Choi I-H, Mang JS, Ko KB. Operating cost reduction of UF membrane filtration process for drinking water treatment attributed to chemical cleaning optimization. Journal of Environmental Management. 2018;**206**:1126-1134

[33] Zhou H, Zhang Z, Wang M, Hu T, Wang Z. Enhancement with physicochemical and biological treatments in the removal of pharmaceutically active compounds during sewage sludge anaerobic digestion processes. Chemical Engineering Journal. 2017;**316**:361-369

[34] Amer H, Moustafa WM, Farghali AA, El Rouby WMA, Khalil WF. Efficient Removal of Cobalt(II) and Strontium(II) Metals from Water using Ethylene Diamine Tetra-acetic Acid Functionalized Graphene Oxide. Zeitschrift für Anorganische und Allgemeine Chemie. 2017;**643**. DOI: 10.1002/zaac.201700318

[35] El Rouby WMA, Al-Ghamdi AA, Abdel-Wahab MS, Jilani A. Sunlight-enhanced catalytic degradation over Ag–CuO nanoparticles thin films prepared by DC/RF sputtering technique. Bulletin of Materials Science. 2018;**41**:58

[36] Farghali AA, Bahgat M, ElRouby WMA, Khedr MH. Decoration of multi-walled carbon nanotubes (MWCNTs) with different ferrite nanoparticles and its use as an adsorbent. Journal of Nanostructure in Chemistry. 2013;**3**:50

[37] El Rouby WM, Farghali AA, Hamdedein A. Microwave synthesis of pure and doped cerium (IV) oxide (CeO_2) nanoparticles for methylene blue degradation. Water Science and Technology. 2016;**74**(10):2325-2336

[38] Farghali AA, El Rouby WM, Hamdedein A. Effect of hydrothermal conditions on microstructures of pure and doped CeO_2 nanoparticles and their photo-catalytic activity: Degradation mechanism and pathway of methylene blue dye. Research on Chemical Intermediates. 2017;**43**(12):7171-7192

[39] Hoffmann N, Gramain JC, Bouas-Laurent H. Photochemistry in organic synthesis. ChemInform. 2009;**40**(23)

[40] El Rouby WMA, Comesaña-Hermo M, Testa-Anta M, Carbó-Argibay E, Salgueiriño V, Pérez-Lorenzo M, Correa-Duarte MA. Au-decorated sodium titanate nanotubes as high-performance selective photocatalysts for pollutant degradation. Journal of Physics D: Applied Physics. 2017;**50**:144002. DOI: 10.1088/1361-6463/aa604c

[41] El Rouby WMA. Selective adsorption and degradation of organic pollutants over Au decorated Co doped titanate nanotubes under simulated solar light irradiation. Journal of the Taiwan Institute of Chemical Engineers. 2018;**88**:201-214. DOI: 10.1016/j.jtice.2018.04.003

Chapter 3

Modified Titanium Dioxide for Photocatalytic Applications

John Moma and Jeffrey Baloyi

Additional information is available at the end of the chapter

http://dx.doi.org/10.5772/intechopen.79374

Abstract

Titanium dioxide (TiO_2) has been widely used as a photocatalyst in many environmental and energy applications due to its efficient photoactivity, high stability, low cost, and safety to the environment and humans. However, its large band gap energy, ca. 3.2 eV limits its absorption of solar radiation to the UV light range which accounts for only about 5% of the solar spectrum. Furthermore, the photocatalytic activity of TiO_2 is also limited by the rapid recombination of the photogenerated electron-hole pairs. When used in water treatment applications, TiO_2 has a poor affinity toward organic pollutants, especially hydrophobic organic pollutants. Several strategies have been employed to reduce its band gap energy, its electron-hole recombination rates as well as enhance its absorption of organic pollutants. In this chapter, we review some of the most recent works that have employed the doping, decoration, and structural modification of TiO_2 particles for applications in photocatalysis. Additionally, we discuss the effectiveness of these dopants and/or modifiers in enhancing TiO_2 photoactivity as well as some perspective on the future of TiO_2 photocatalysis.

Keywords: titanium dioxide, photocatalysis, environmental pollution, modification

1. Introduction

The rapid growth of global population as well as industrialization has led to a concomitant increase in environmental pollution. This has very negative effects on natural elements that are vital for life on earth such as air and water. It becomes very crucial therefore to find sustainable ways to mitigate pollution in order to provide a clean and safe environment for humans. Photocatalysis has attracted worldwide interest due to its potential to use solar energy not only to solve environmental problems but also provide a renewable and sustainable energy source. An efficient photocatalyst converts solar energy into chemical energy which can be used for environmental and energy applications such as water treatment, air purification, self-cleaning surfaces, hydrogen production by water cleavage and CO_2 conversion to hydrocarbon fuels.

Research in the development of efficient photocatalytic materials has seen significant progress in the last 2 decades with a large number of research papers published every year. Improvements in the performance of photocatalytic materials have been largely correlated with advances in nanotechnology. Of many materials that have been studied for photocatalysis, titanium dioxide (TiO_2; titania) has been extensively researched because it possesses may merits such as high photocatalytic activity, excellent physical and chemical stability, low cost, non-corrosive, nontoxicity and high availability [1–4]. The photocatalytic activity of titania depends on its phase. It exists in three crystalline phases; the anatase, rutile and brookite. The anatase phase is metastable and has a higher photocatalytic activity, while the rutile phase is more chemically stable but less active. Some titania with a mixture of both anatase and rutile phases exhibit higher activities compared to pure anatase and rutile phases [5–7]. When titania is irradiated with light of sufficient energy, electrons from the valence band are promoted to the conduction band, leaving an electron deficiency or hole, h^+, in the valence band and an excess of negative charge in the conduction band. The free electrons in the conduction band are good reducing agents while the resultant holes in the valence band are strong oxidizing agents and can both participate in redox reactions.

Titania however suffers from a number of drawbacks that limit its practical applications in photocatalysis. Firstly, the photogenerated electrons and holes coexist in the titania particle and the probability of their recombination is high. This leads to low rates of the desired chemical transformations with respect to the absorbed light energy [8, 9]. The relatively large band gap energy (~ 3.2 eV) requires ultraviolet light for photoactivation, resulting in a very low efficiency in utilizing solar light. UV light accounts for only about 5% of the solar spectrum compared to visible light (45%) [1, 10]. In addition to these, because titania is non-porous and has a polar surface, it exhibits low absorption ability for non-polar organic pollutants [10–13]. There is also the challenge to recover nano-sized titania particles from treated water in regards to both economic and safety concern [14]. The TiO_2 nanoparticles also suffer from aggregation and agglomeration which affect the photoactivity as well as light absorption [15–18]. Several strategies have been employed in the open literature to overcome these drawbacks. These strategies aim at extending the wavelength of photoactivation of TiO_2 into the visible region of the spectrum thereby increasing the utilization of solar energy; preventing the electron/hole pair recombination and thus allowing more charge carriers to successfully diffuse to the surface; increasing the absorption affinity of TiO_2 towards organic pollutants as well as preventing the aggregation and agglomeration of the nano-titania particles while easing their recovery from treated water. Several reviews have been published in recent years on the development of strategies to eliminate the limitations of titania photocatalysis [1, 19–25]. Most of these however focus on pollutant removal from wastewater, water splitting for hydrogen production, CO_2 conversion and reaction mechanisms [1, 21, 25–31]. In this chapter, we review some of the latest publications mainly covering the last 5 years, on strategies that have been researched to overcome the limitations of TiO_2 for general photocatalytic applications and the level of success that these strategies have been able to achieve. Based on the current level of research in this field, we also present some perspectives on the future of modified TiO_2 photocatalysis.

2. Modification of TiO_2 photocatalysts

A large number of research works have been published on TiO_2 modification to enhance its photocatalytic properties. These modifications have been done in many different ways which include metal and non-metal doping, dye sensitization, surface modification, fabrication of composites with other materials and immobilization and stabilization on support structures. The properties of modified TiO_2 are always intrinsically different from the pure TiO_2 with regards to light absorption, charge separation, adsorption of organic pollutants, stabilization of the TiO_2 particles and ease of separation of TiO_2 particles.

2.1. Metal doping

Metal doping has been extensively used to advance efforts at developing modified TiO_2 photocatalysts to operate efficiently under visible light. The photoactivity of metal-doped TiO_2 photocatalysts depends to a large extent on the nature of the dopant ion and its nature, its level, the method used in the doping, the type of TiO_2 used as well as the reaction for which the catalyst is used and the reaction conditions [32]. The mechanism of the lowering of the band gap energy of TiO_2 with metal doping is shown in **Figure 1**. It is believed that doping TiO_2 with metals results in an overlap of the Ti 3d orbitals with the d levels of the metals causing a shift in the absorption spectrum to longer wavelengths which in turn favours the use of visible light to photoactivate the TiO_2.

Doping of TiO_2 nanoparticles with Li, Na, Mg, Fe and Co by high energy ball milling with the metal nitrates was found to widen the TiO_2 visible light response range. In the Na-doped sample, Ti existed as both Ti^{4+} and Ti^{3+} and the conversion between Ti^{4+} and Ti^{3+} was found to prevent the recombination of electrons and (e^-) and holes (h^+). The metal ion doping promoted crystal phase transformations that generated electrons (e^-) and holes (h^+) [33]. Mesoporous TiO_2 prepared by sol gel technique and doped with different levels of Pt (1–5 wt% nominal loading) resulted in a high surface area TiO_2 with an enhanced catalytic performance in photocatalytic water splitting for the Pt-doped samples. The 2.5 wt%Pt-TiO_2 had showed the optimum catalytic performance and a reduction in the TiO_2 band gap energy from 3.00 to 2.34 eV with an enhanced electron storage capacity, leading to a minimization of the electron-hole recombination rate [34]. Noble metal nanoparticles such as Ag [35], Pt [34], Pd [36], Rh [37] and Au [38] have also been used to modify TiO_2 for photocatalysis and have been reported to efficiently hinder electron-hole recombination due to the resulting Schottky barrier at the metal-TiO_2 interface. The noble metal nanoparticles act as a mediator in storing and transporting photogenerated electrons from the surface of TiO_2 to an acceptor. The photocatalytic activity increases as the charge carriers recombination rate is decreased.

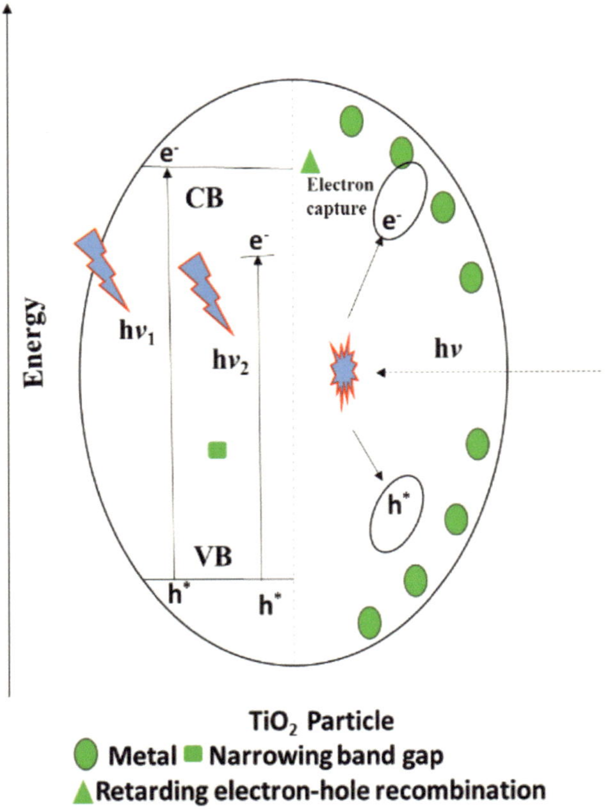

Figure 1. Band-gap lowering mechanism of metal-doped TiO_2.

In a recent review by Low et al. [21] the deposition of Au onto TiO_2 surface is reported to result in electron transfer from photo-excited Au particles (> 420 nm) to the conduction band of TiO_2, which showed a decrease in their absorption band (~550 nm) and the band was recovered by the addition of electron donors such as Fe^{2+} and alcohols. Zhang et al. [39] reported that the visible light activity of coupled Au/TiO_2 can be ascribed to the electric field enhancement near the metal nanoparticles. Moreover, numerous researchers coupled Au and Ag nanoparticles onto TiO_2 surface to use their properties of localized surface plasmonic resonance (LSPR) in photocatalysis [40]. Wang et al. [41] and Hu et al. [42] reported an improved photocatalytic performance due to the Pt nanoparticle which increased the electron transfer rate to the oxidant. It was observed that photocatalytic sacrificial hydrogen generation was influenced by several parameters such as platinum loading (wt%) on TiO_2, solution pH, and light (UV, visible and solar) intensities [43]. Moreover, complete discoloration and dye mineralization were achieved using Pt/TiO_2 as catalyst; the results were attributed to the higher Pt content of the photocatalyst prepared with the highest deposition time. For Pt-TiO_2 catalysts the best discoloration and dye mineralization were obtained over the catalyst prepared by photochemical deposition method and using 120 min of deposition time in the synthesis. These results may be due to the higher Pt content of the photocatalyst prepared with the highest deposition time.

Haung et al. [44] prepared Pt/TiO_2 nanoparticles from TiO_2 prepared at various hydrolysis pH values and found that the phase of TiO_2 obtained depended largely on the hydrolysis pH. The anatase/rutile intersection of a Pt/TiO_2 sample had a lower recombination rate compared to the anatase phase of Pt/TiO_2 due to the longer recombination pathway. Though, the Pt/TiO_2 anatase phase showed better degradation efficiency than the Pt/TiO_2 anatase/rutile intersection. The decrease in the anatase composition of TiO_2, and the decrease in the composition of TiO_2 resulted in the degradation rate decrease, suggesting that anatase composition in the Pt/TiO_2 system played a crucial role of increasing the photocatalytic degradation of Acid Red 1 dye.

Liu et al. [45] prepared the palladium doped TiO_2 (Pd-TiO_2) photocatalyst using chemical reduction method and tested it the photocatalytic degradation of organic pollutant. It was found that the TiO_2 grain size was reduced while the specific surface area increased and the absorption of ultraviolet light also enhanced after using chemical reduction method, however, all these changes had no effect on degradation of organic pollutant. But the degradation was significantly improved due to the deposition of Pd nanoparticles; the Pd/TiO_2 organic pollutant degradation was 7.3 times higher compared to TiO_2 (P25).

Repouse et al. [46] prepared a series of noble metal promoted TiO_2 (P25) by wet impregnation and found that the dispersion of the small metal crystallites on TiO_2 did not affect the optical band gap of TiO_2. The Pt-promoted catalyst exhibited the highest photocatalytic efficiency in the degradation of bisphenol A under solar irradiation. They also found the presence of humic acid to considerably improve the reaction rate of Rh/TiO_2 but had a clearly adverse effect with P25 TiO_2 photocatalyst. Fluorescence data revealed that humic acid is capable of photosensitizing the Rh/TiO_2 catalyst.

Indium-doped TiO_2 have recently been used for photocatalytic reduction of CO_2 [47]. Indium doping resulted in an increase in surface area because of suppression of TiO_2 particle growth during the TiO_2 synthesis. The light absorption ability of the In-TiO_2 was enhanced due to the introduction of the impurity level below the conduction band level of the TiO_2. The photocatalytic CO_2 reduction activity of the In-TiO_2 was about 8 time that of pure TiO_2 as a consequence of the high surface area and extended light absorption range.

The doping of TiO_2 with transition metals such as Cr [48], Co [48], Fe [48–50], Ni [48, 51], Mn [48, 52], V [53], Cu [54], Ni [51] and Zn [55], has been studied by different research groups. Numerous studies reported that doping of TiO_2 with transition metals improve the photocatalytic activity, attributable to a change in the electronic structure resulting in the absorption region being shifted from UV to visible light. The shift results from charge-transfer transition between the d electrons of the transition metals and the conduct or valence band of TiO_2 nanoparticles. Inturi et al. [48] compared the doping of TiO_2 nanoparticles with Cr, Fe, V, Mn, Mo, Ce, Co, Cu, Ni, Y and Zr and it was found that Cr, Fe and V showed improved conversions in the visible region while, the incorporation of the other transition metals (Mn, Mo, Ce, Co, Cu, Ni, Y and Zr) exhibited an inhibition effect on the photocatalytic activity. The Cr-doped TiO_2 demonstrated a superior catalytic performance and the rate constant was found to be approximately 8–19 times higher than the rest of the metal doped catalysts. It was reported that the reduction peaks in Cr-doped TiO_2 shifted to much lower temperatures, due to the increase in the reduction potential of titania and chromium. Therefore, the higher photocatalytic efficiency of Cr/TiO_2 in the visible light can be attributed to strong interaction (formation of Cr-O-Ti bonds). Fe-doped TiO_2 nanoparticles were used in the visible light degradation of para-nitrophenol and it was found that the Fe-dopant concentration was crucially important in determining the activity of the catalyst. The maximum degradation rate of para-nitrophenol observed was 92% in 5 h when the Fe(3+) molar concentration was 0.05 mol%, without addition of any oxidizing reagents. The excellent photocatalytic activity was as a result of an increase in the threshold wavelength response as well as maximum separation of photogenerated charge carriers [49]. On the other hand, Fe-doped TiO_2 evaluated for solar photocatalytic activity for the degradation of humic acid showed a retardation effect for the doped catalysts compared to the bare TiO_2 specimens, which could be attributed to surface complexation reactions rather than the reactions taking place in aqueous medium. The faster removal rates attained by using bare TiO_2 could be regarded as substrate specific rather than being related to the inefficient visible light activated catalytic performance [50]. Ola et al. [56] reported that the properties of V doped TiO_2 were tuned towards visible light because of the substitution of the Ti^{4+} by V^{4+} or V^{5+} ions since the V^{4+} is centred at 770 nm while the absorption band of V^{5+} is lower than 570 nm. Moradi et al. [57] obtained high photocatalytic activity of Fe doped TiO_2 and studied the effects of Fe^{3+} doping content on the band gap and size of the nanoparticles. It was found that the increase in the doping content decreased the band gap energy and particle size from 3.3 eV and 13 nm for bare TiO_2 to 2.9 eV and 5 nm for Fe_{10}-TiO_2, respectively.

The rare earth metals doped TiO_2 catalyst also have good electron trapping properties which can result in a stronger absorption edge shift towards longer wavelength, obtaining narrow band gap. Bethanabotla et al. [58] carried out a comprehensive study on the rare earth doping into a TiO_2 and found that the rare earth dopants improved the aqueous-phase photodegradation of phenol at low loadings under simulated solar irradiation, with improvements varying by catalyst composition. Differences in defect chemistry on key kinetic steps were given as the explanation for the enhanced performance of the rare earth doped samples compared to pure titania. Reszczyńska et al. [59] prepared a series of Y^{3+}, Pr^{3+}, Er^{3+} and Eu^{3+} modified TiO_2 nanoparticles photocatalysts and results demonstrate that the incorporation of RE^{3+} ions into TiO_2 nanoparticles resulted in blue shift of absorption edges of TiO_2 nanoparticles and could be ascribed to movement of conduction band edge above the first excited state of RE^{3+}. Moreover, incorporated RE^{3+} ions at the first excited state interact with the electrons of the conduction band of TiO_2, resulting in a higher energy transfer from the TiO_2 to RE^{3+} ions. But observed blue shift could be also attributed to decrease in crystallite size of RE^{3+}–TiO_2 in comparison to TiO_2. The Y^{3+}, Pr^{3+}, Er^{3+} and Eu^{3+} modified TiO_2 nanoparticles exhibited higher activity under visible light irradiation compared to pure P25 TiO_2 and can be excited under visible light in the range from 420 to 450 nm. In a similar work on rare earths (Er, Yb, Ho, Tb, Gd and Pr) titania nanotubes (RE-NTs), [60] the RE^{3+} species were found to be located at the crystal boundaries rather than inside the TiO_2 unit cell and an observed excitation into the TiO_2 absorption band with resulting RE^{3+} emission confirmed energy migration between the TiO_2 matrix and RE^{3+}. The presence of the rare earth component was found to reduce recombination of the electrons and holes successfully by catching them and also by promoting their rapid development along the surface of TiO_2 nanoparticles. Lanthanide ions doping did not impact the energy gap of TiO_2 nanoparticles, however this enhanced the light absorption of catalyst. The surface range of TiO_2 nanoparticles generally increases by La^{3+} particle doping by diminishing the crystallite size and accordingly, the doped TiO_2 nanoparticle displayed higher adsorption capacity. Based on theoretical calculations, it was proposed that during the electrochemical process, new Ho-f states and surface vacancies were formed and may reduce the photon excitation energy from the valence to the conduction band under visible light irradiation. The photocatalytic activity under visible light irradiation was attributed not to ·OH but to other forms of reactive oxygen species ($O_2^{\cdot -}$, HO_2, H_2O_2).

2.2. Non-metal doping

TiO_2 nanoparticles have been comprehensively doped at the O sites with non-metals such as C [61], B [62], I [63], F [64], S [65], and N [66]. Non-metal dopants are reported to be more appropriate for the extension of the photocatalytic activity of TiO_2 into visible region compared to metal dopant [67, 68]. This can be ascribed to the impurity states which are near the valence band edge, however, they do not act as charge carriers, and their role as recombination centres might be minimized [53]. As shown in **Figure 2**, the mixing of the p states of

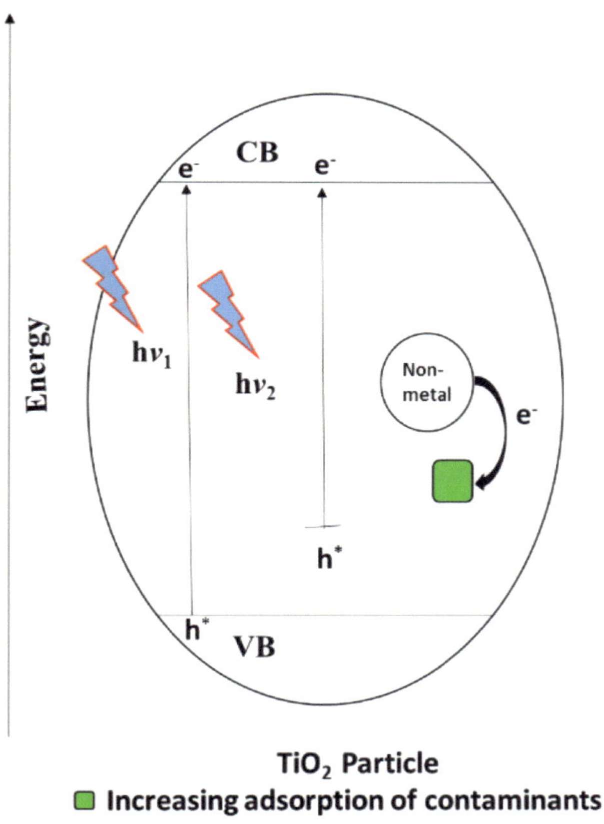

Figure 2. Band-gap energy narrowing mechanism for non-metal-doped TiO_2.

the doped non-metal with the O2p states shifts the valence band edge upward and narrows the band-gap energy of the doped TiO_2 photocatalyst. The nitrogen and carbon doped TiO_2 nanoparticles has been reported to exhibit greater photocatalytic activity under visible light irradiation compared to other non-metal dopants.

N-doped TiO_2 (N-TiO_2) appears to be the most efficient and extensively investigated photocatalyst for non-metal doping. Zeng et al. [69] reported the preparation of a highly active modified N-TiO_2 nanoparticle via a novel modular calcination method. The excellent photocatalytic performance of the photocatalyst was ascribed to excellent crystallinity, strong light harvesting and fast separation of photogenerated carriers. Moreover, the enhancement of charge separation was attributed to the formation of paramagnetic [O-Ti^{4+}-N^{2-}-Ti^{4+}-V_O] cluster. The surface oxygen vacancy induced by vacuum treatment trapped electron and promoted to generate super oxygen anion radical which was a necessary active species in photocatalytic process. Phongamwong et al. [70] investigated the photocatalytic activity of CO_2 reduction under visible light over modified N-TiO_2 photocatalyst and they have found that the band gap of N-TiO_2 photocatalyst slightly decreases with increasing N content. In addition, the sub-band energies related to the impurity energy level were observed in the N-TiO_2 photocatalyst because of the interstitial N species and the sub-band gap energies were found to have decreased from 2.18 eV with 10 wt% N-TiO_2 photocatalyst. In contrast, the replacement of O by N is difficult because of the radius of N (17.1 nm) being higher compared to O (14 nm) and the electroneutrality can be maintained by oxygen vacancies, that are provided by replacement of three oxygen vacancies by two nitrogen atom [71]. N-TiO_2 photocatalyst reduces the oxygen energy vacancies from 4.2 to 0.6 eV, suggesting that N favors the formation of oxygen vacancies [72].

In contrast, O atoms (14 nm) could be substituted easily by F atoms (13.3 nm) because of their similar ionic radius [73]. Yu et al. [64] reported that the F-doped TiO_2 (F-TiO_2) is able to absorb visible light due to the high-density states that were evaluated to be below the maxima valence band, although there was no shift in the band edge of TiO_2. Samsudin et al. found a synergistic effect between fluorine and hydrogen in hydrogenated F-doped TiO_2 which enabled light absorption in UV, visible and infrared light illumination with enhanced electrons and holes separation. Surface vacancies and Ti^{3+} centres of the hydrogenated F-doped catalyst coupled with enhanced surface hydrophilicity facilitated the production of surface-bound and free hydroxyl radicals. Species present on the surface of the catalyst triggered the formation of new Ti^{3+} occupied states under the conduction band of the hydrogenated F-doped TiO_2, thus narrowing the band gap energy [73]. Enhanced photocatalytic performance of N-doped TiO_2 over pure TiO_2 has also been ascribed to efficient separation of electron-hole pairs as well as an increased creation of surface radicals such as hydroxyl The band gap can also be narrowed by doping TiO_2 with S, since replacement of S into TiO_2 can be performed easily due to larger radius of S atoms (18 nm) compared to O atoms (14 nm). S incorporation in TiO_2 has been reported to change the lattice spacing of the TiO_2 with a reduction in the band gap width from 3.2 to 1.7 eV allowing for higher photocatalytic activity [74]. N, S and C co-doped TiO_2 samples photocatalytic reduction of Cr(IV) showed that the co-doping and calcination played an important role in the microstructure and photocatalytic activity of the catalysts. The co-doped samples calcined at 500°C showed the highest activities ascribed to the synergistic effect in enhancing crystallization of anatase and (N, S and C) co-doping. The carbon doped TiO_2 (C-TiO_2) is reported to be more active than N-TiO_2, therefore, C-TiO_2 has received special attention [75]. Noorimotlagh et al. [76] investigated the photocatalytic removal of nonylphenol (NP) compound using visible light active C-TiO_2 with anatase/rutile. It was found that the doping of C into TiO_2 lattice may enhance the visible light utilization and affect the structural properties of the as-synthesized photocatalysts. Moreover, it was reported that after C doping and changing the calcination temperature, the band gap was narrowed from 3.17 to 2.72 eV and from 2.72 to 2.66 eV, respectively. Ji et al. [61] reported the preparation of C-TiO_2 with a diameter of around 200 nm and the tube wall was composed of anatase TiO_2, amorphous carbon, crystalline carbon and carbon element doping into the lattice of TiO_2. The C-TiO_2 nanotubes exhibited much better performance in photocatalytic activity than bare TiO_2 under UV and visible light. The obtained results were ascribed to the C doping, which narrowed the band gap energy of TiO_2, extended the visible light adsorption toward longer wavelength and hindered charge recombination.

2.3. Co-doping and tri-doping

Although single metal doped and non-metal doped TiO_2 have exhibited excellent performance in decreasing the electrons and holes recombination, but they suffer from thermal stability and losing a number of dopants during catalyst preparation process [77]. Therefore, co-doping of two kinds of atoms into TiO_2 has recently attracted much interest [78]. The electronic structure of TiO_2 can be altered by co-doping on TiO_2 by formation of new doping levels inside its band gap. Abdullah et al. [77] reported that the doping levels situated within the band gap of TiO_2 can either accept photogenerated electrons from TiO_2 valence band or absorb photons with longer wavelengths. Therefore, suggesting that the TiO_2 absorption range can be expanded.

Zang et al. [79] evaluated the photocatalytic degradation of atrazine under UV and visible light irradiation by N,F-codoped TiO_2 nanowires and nanoparticles in aqueous phase. It was found that photocatalytic degradation of atrazine was higher in the presence of N,F-codoped TiO_2 nanowires than that of N,F-codoped TiO_2 nanoparticles. The higher photocatalytic performance in the presence of N,F-codoped TiO_2 nanowires was attributed to the higher charge carrier mobility and lower carrier recombination rate. Moreover, the speed of electron diffusion across nanoparticle intersections is several orders of magnitude smaller compared to that of nanowire because of frequent electron trapping at the intersections of nanoparticles and increasing the recombination of separated charges before they reach the TiO_2 nanoparticles surface. Park et al. [80] showed the best performance for novel Cu/N-doped TiO_2 photoelectrodes for dye-sensitized solar cells. It was found that the Cu/N-doped TiO_2 nanoparticles provided higher surface area, active charge transfer and decreased charge recombination. Moreover, the addition of suitable content of Cu- to N-doped TiO_2 electrode effectively inhibited the growth of TiO_2 nanoparticles and improved the optical response of the photoelectrode under visible light irradiation. Chatzitakis et al. [81] studied the photoelectrochemical properties of C, N, F codoped TiO_2 nanotubes. It was found that increasing surface area is not followed by increase in the photoconversion efficiency, but rather that an optimal balance between electroactive surface area and charge carrier concentration occurs.

Zhao et al. [82] investigated the photocatalytic H_2 evolution performance of Ir-C-N tridoped TiO_2 under UV-visible light irradiation. The photocatalytic activity of TiO_2 nanoparticles was reported to be improved by Ir-C-N tridoped TiO_2 under UV-visible light, due the synergistic effect between Ir, C and N on the electron structure of TiO_2. It was found that Ir existed as Ir^{4+} by substituting Ti in the lattice of TiO_2 nanoparticles, whereas the C and N were also incorporated into the surface of TiO_2 nanoparticles in interstitial mode. The absorption of TiO_2 nanoparticles was expanded into the visible light region and the band gap was narrowed to ~3.0 eV, resulting in improved photocatalytic H_2 evolution under UV-visible light irradiation. Tan et al. [83] investigated the photocatalytic degradation of methylene blue by W–Bi–S-tridoped TiO_2 nanoparticles. It was found that the absorption edge of TiO_2 was expanded into visible-light region after doping with W, Bi and S and the catalytst showed the best photocatalytic activity, than that of TiO_2, S-TiO_2, W–S–TiO_2 and Bi–S–TiO_2. This might be attributed to the synergistic effect of W, Bi and S.

2.4. Nano-structured TiO_2

Amongst the various strategies that have been used to enhance TiO_2 photocatalytic activity, improvement of morphology, crystal structure and surface area have also been considered important and widely investigated approach to achieve better photocatalytic performance. The nanotitania crystallinity can simply be enhanced by optimizing the annealing temperature. However, the stability of the structure and geometries have to be considered when annealing [84]. For the nanotitania morphology and surface area, various ordered structures have been studied. TiO_2 nanotubes [85, 86], nanowires [79], nanospheres [87], etc. Tang et al. fabricated monodisperse mesoporous anatase TiO_2 nanospheres using a template material and found the resulting catalysts to show high photocatalytic degradation efficiency and selectivity towards different target dye molecules and could be readily separated from a slurry system after photocatalytic reaction [87]. Anodic TiO_2 nanotubes have been reported to allow a high control over the separation of photogenerated charge carriers in photocatalytic reactions. The nanotube array has as key advantage the fact that nanotube modifications can be embedded site specifically into the tube wall or at defined locations along the tube wall. This allows for engineering of reaction sites giving rise to enhanced photocatalytic efficiencies and selectivities [88].

2.5. Nanocarbon modified TiO_2

The design and preparation of graphene-based composites containing metal oxides and metal nanoparticles have attracted attention for photocatalytic performances. For example, Tan et al. [89] prepared a novel graphene oxide-doped-oxygen-rich TiO_2 (GO–OTiO_2) hybrid heterostructure and evaluated its activity for photoreduction of CO_2 under the irradiation of low-power energy-saving daylight bulbs. It was found that the photostability of O_2–TiO_2 was significantly improved by the addition of GO, at which the resulting hybrid composite retained a high reactivity. The photoactivity attained was about 1.6 and 14.0 folds higher than that of bare O_2–TiO_2 and the commercial Degussa P25, respectively. This high photocatalytic performance of GO–OTiO_2 was attributed to the synergistic effect of the visible-light-responsiveness of O_2–TiO_2 and an enhanced separation and transfer of photogenerated charge carriers at the intimate interface of GO–OTiO_2 heterojunctions. This study is reported to have opened up new possibilities in the development of novel, next generation heterojunction photocatalysts for energy and environmental related applications. Lin et al. [90] also investigated photoreduction of CO_2 with H_2O vapor in the gas-phase under the irradiation of a Xe lamp using TiO_2/nitrogen (N) doped reduced graphene oxide (TiO_2/NrGO) nanocomposites. They found that the quantity and configuration of N dopant in the TiO_2/NrGO nanocomposites strongly influenced the photocatalytic efficiency, and the highest catalytic activity was observed for TiO_2/NrGO nanocomposites with the highest N doping content. Moreover, modified TiO_2/rGO demonstrated a synergistic effect, enhancing CO_2 adsorption on the catalyst surface and promoting photogenerated electron transfer that resulted in a higher CO_2 photoreduction rate of TiO_2/NrGO. Qu et al. [91] prepared the graphene quantum dots (GQDs) with high quantum yield (about 23.6% at an excitation wavelength of 320 nm) and GQDs/TiO_2 nanotubes (GQDs/TiO_2 nanoparticles) nanocomposites and the photocatalytic activity was tested towards the degradation of methyl orange. It was found that the GQDs deposited on TiO_2 nanoparticles can expand the visible light absorption of TiO_2 nanoparticles and enhance the activity on photocatalytic degradation of methyl orange under UV-vis light irradiation (λ = 380–780 nm). Furthermore, the photocatalytic activity of GQDs/TiO_2 nanoparticles was approximately 2.7 times as higher than that of bare TiO_2 nanoparticles. Tian et al. [92] reported the preparation of N, S co-doped graphene quantum dots (N, S-GQDs)-reduced graphene oxide- (rGO)-TiO_2 nanotubes (TiO_2NT) nanocomposites for photodegradation of methyl orange under visible light irradiation. It was found that the S-GQDs+rGO + TiO_2 nanocomposites simultaneously showed an extended photoresponse range, improved charge separation and transportation properties. Moreover, the apparent rate constant of N, S-GQDs+rGO + TiO_2NT is 1.8 and 16.3 times higher compared to rGO + TiO_2NT and pure TiO_2NT, respectively. Suggesting that GQDs can improve the utilization of solar light for energy conversion and environmental therapy.

2.6. Immobilized TiO_2

Another drawback of TiO_2 nanoparticles mentioned above is the formation of uniform suspension in water which makes its recovery difficult, therefore hindering the application of photocatalytic in an industrial scale. As a result, many studies have attempted the modification of TiO_2 nanoparticles on support materials such as clays [93, 94] quartz [95], stainless steel [96], etc. Clays have been reported to be a significant support material for TiO_2 nanoparticles because of their layered morphology, chemical as well as mechanical stability, cation exchange capacity, non-toxic nature, low cost and availability. Therefore, TiO_2/clay nanocomposites have attracted much attention for application in both water and air purification and have been prepared by numerous researchers. Belver et al. [97] investigated the removal of atrazine under solar light using a novel W-TiO_2/clay photocatalysts. It was found that the photocatalytic activity of W-TiO_2/clay catalyst exhibited higher photocatalytic performance than that of an un-doped TiO_2/clay, which was explained by the presence of W ions in the TiO_2 nanostructure. The substitution of Ti ions with W resulted in the increase of its crystal size and the distortion of its lattice and moderately narrower band gap of photocatalysts. Mishra et al. [98] reported the preparation of TiO_2/clay nanocomposites for photocatalytic degradation of VOC and dye. They found that the photocatalytic performance of TiO_2/clay nanocomposites is highly dependent on the clay texture (as 2:1 clays show highest activity than 1:1) apart from their surface area and porosity. Moreover, the reactions involving TiO_2/Clay photocatalyst were fast with rate constant of 0.02886 and 0.04600 min^{-1} for dye and VOC respectively than the other nanocomposites.

3. Conclusions

In this chapter, we have given an overview of the development of modified TiO_2 catalysts and its future prospects from a scientific point of view. We note that the field has experienced major advances in the last 5 years especially in the area of modifying TiO_2 with carbon nanomaterials. Based on the literature we have covered here, we believe that there is still quite a lot that can be achieved in improving the performance of TiO_2 catalysts for photocatalytic applications.

Conflict of interest

There are no conflicts of interest to declare.

Author details

John Moma[1]* and Jeffrey Baloyi[1,2]

*Address all correspondence to: john.moma@wits.ac.za

1 School of Chemistry, University of the Witwatersrand, Johannesburg, South Africa

2 Advanced Materials Division, Mintek, Malibongwe Drive, Randburg, South Africa

References

[1] Dong H, Zeng G, Tang L, Fan C, Zhang C, He X. An overview on limitations of TiO_2-based particles for photocatalytic degradation of organic pollutants and the corresponding countermeasures. Water Research. 2015;**79**:128-146

[2] Jiang L, Wang Y, Feng C. Application of photocatalytic technology in environmental safety. Procedia Engineering. 2012;**45**:993-997

[3] Tasbihi M, Călin I, Šuligoj A, Fanetti M, Lavrenčič Štangar U. Photocatalytic degradation of gaseous toluene by using TiO_2 nanoparticles immobilized on fiberglass cloth. Journal of Photochemistry and Photobiology A: Chemistry. 2017;**336**:89-97

[4] Baloyi J, Seadira T, Raphulu M, Ochieng A. Preparation, characterization and growth mechanism of dandelion-like TiO_2 nanostructures and their application in photocatalysis towards reduction of Cr (VI). Materials Today: Proceedings. 2015;**2**(7):3973-3987

[5] Valencia S, Marín JM, Restrepo G. Study of the bandgap of synthesized titanium dioxide nanoparticules using the sol-gel method and a hydrothermal treatment. The Open Materials Science Journal. 2010;**4**:9-14

[6] Jing L, Li S, Song S, Xue L, Fu H. Investigation on the electron transfer between anatase and rutile in nano-sized TiO_2 by means of surface photovoltage technique and its effects on the photocatalytic activity. Solar Energy Materials and Solar Cells. 2008;**92**(9):1030-1036

[7] Ohno T, Sarukawa K, Tokieda K, Matsumura M. Morphology of a TiO_2 photocatalyst (Degussa, P-25) consisting of anatase and rutile crystalline phases. Journal of Catalysis. 2001;**203**(1):82-86

[8] Etacheri V, Di Valentin C, Schneider J, Bahnemann D, Pillai SC. Visible-light activation of TiO_2 photocatalysts: Advances in theory and experiments. Journal of Photochemistry and Photobiology C: Photochemistry Reviews. 2015;**25**:1-29

[9] Riboni F, Bettini LG, Bahnemann DW, Selli E. WO3–TiO_2 vs. TiO_2 photocatalysts: Effect of the W precursor and amount on the photocatalytic activity of mixed oxides. Catalysis Today. 2013;**209**:28-34

[10] Szczepanik B. Photocatalytic degradation of organic contaminants over clay-TiO_2 nanocomposites: A review. Applied Clay Science. 2017;**141**:227-239

[11] Bhattacharyya A, Kawi S, Ray MB. Photocatalytic degradation of orange II by TiO_2 catalysts supported on adsorbents. Catalysis Today. 2004;**98**(3):431-439

[12] Torimoto T, Okawa Y, Takeda N, Yoneyama H. Effect of activated carbon content in TiO_2-loaded activated carbon on photodegradation behaviors of dichloromethane. Journal of Photochemistry and Photobiology A: Chemistry. 1997;**103**(1):153-157

[13] Lepore GP, Persaud L, Langford CH. Supporting titanium dioxide photocatalysts on silica gel and hydrophobically modified silica gel. Journal of Photochemistry and Photobiology A: Chemistry. 1996;**98**(1):103-111

[14] Tang Y, Zhang G, Liu C, Luo S, Xu X, Chen L, et al. Magnetic TiO_2-graphene composite as a high-performance and recyclable platform for efficient photocatalytic removal of herbicides from water. Journal of Hazardous Materials. 2013;**252**(253):115-122

[15] Pellegrino F, Pellutiè L, Sordello F, Minero C, Ortel E, Hodoroaba V-D, et al. Influence of agglomeration and aggregation on the photocatalytic activity of TiO_2 nanoparticles. Applied Catalysis B: Environmental. 2017;**216**:80-87

[16] Melcher J, Barth N, Schilde C, Kwade A, Bahnemann D. Influence of TiO_2 agglomerate and aggregate sizes on photocatalytic activity. Journal of Materials Science. 2017;**52**(2): 1047-1056

[17] De Temmerman P-J, Verleysen E, Lammertyn J, Mast J. Semi-automatic size measurement of primary particles in aggregated nanomaterials by transmission electron microscopy. Powder Technology. 2014;**261**:191-200

[18] Ivanova I, Mendive CB, Bahnemann D. The role of nanoparticulate agglomerates in TiO_2 photocatalysis: Degradation of oxalic acid. Journal of Nanoparticle Research. 2016;**18**(7):187

[19] Ni M, Leung MKH, Leung DYC, Sumathy K. A review and recent developments in photocatalytic water-splitting using TiO_2 for hydrogen production. Renewable and Sustainable Energy Reviews. 2007;**11**(3):401-425

[20] Schneider J, Matsuoka M, Takeuchi M, Zhang J, Horiuchi Y, Ampo M, et al. Understanding TiO_2 photocatalysis: Mechanisms and materials. Chemical Reviews. 2014;**114**(19):9919-9986

[21] Low J, Cheng B, Yu J. Surface modification and enhanced photocatalytic CO_2 reduction performance of TiO_2: A review. Applied Surface Science. 2017;**392**:658-686

[22] Verbruggen SW. TiO_2 photocatalysis for the degradation of pollutants in gas phase: From morphological design to plasmonic enhancement. Journal of Photochemistry and Photobiology C: Photochemistry Reviews. 2015;**24**:64-82

[23] Liu N, Chen X, Zhang J, Schwank JW. A review on TiO_2-based nanotubes synthesized via hydrothermal method: Formation mechanism, structure modification, and photocatalytic applications. Catalysis Today. 2014;**225**:34-51

[24] Wen J, Li X, Li W, Fang Y, Xie J, Xu Y. Photocatalysis fundamentals and surface modification of TiO_2 nanomaterials. Chinese Journal of Catalysis. 2015;**36**(12):2049-2070

[25] Daghrir R, Drogui P, Robert D. Modified TiO_2 for environmental photocatalytic applications: A review. Industrial & Engineering Chemistry Research. 2013;**52**(10):3581-3599

[26] Pelaez M, Nolan NT, Pillai SC, Seery MK, Falaras P, Kontos AG, et al. A review on the visible light active titanium dioxide photocatalysts for environmental applications. Applied Catalysis B: Environmental. 2012;**125**:331-349

[27] Ahmad H, Kamarudin SK, Minggu LJ, Kassim M. Hydrogen from photo-catalytic water splitting process: A review. Renewable and Sustainable Energy Reviews. 2015;**43**:599-610

[28] Liu L, Li Y. Understanding the reaction mechanism of photocatalytic reduction of CO_2 with H_2O on TiO_2-based photocatalysts: A review. Aerosol and air quality research. 2014; **14**(2):453-469

[29] Morales-Torres S, Pastrana-Martínez LM, Figueiredo JL, Faria JL, Silva AMT. Design of graphene-based TiO_2 photocatalysts—A review. Environmental Science and Pollution Research. 2012;**19**(9):3676-3687

[30] Zangeneh H, Zinatizadeh AAL, Habibi M, Akia M, Hasnain Isa M. Photocatalytic oxidation of organic dyes and pollutants in wastewater using different modified titanium dioxides: A comparative review. Journal of Industrial and Engineering Chemistry. 2015; **26**:1-36

[31] Kumar SG, Devi LG. Review on modified TiO_2 photocatalysis under UV/visible light: Selected results and related mechanisms on interfacial charge carrier transfer dynamics. The Journal of Physical Chemistry A. 2011;**115**(46):13211-13241

[32] Kuvarega AT, Mamba BB. TiO_2-based photocatalysis: Toward visible light-responsive photocatalysts through doping and fabrication of carbon-based nanocomposites. Critical Reviews in Solid State and Materials Sciences. 2017;**42**(4):295-346

[33] Zhao Q, Wang M, Yang H, Shi D, Wang Y. Preparation, characterization and the antimicrobial properties of metal ion-doped TiO_2 nano-powders. Ceramics International. 2018;**44**(5):5145-5154

[34] Guayaquil-Sosa JF, Serrano-Rosales B, Valadés-Pelayo PJ, de Lasa H. Photocatalytic hydrogen production using mesoporous TiO_2 doped with Pt. Applied Catalysis B: Environmental. 2017;**211**:337-348

[35] Hossain MA, Elias M, Sarker DR, Diba ZR, Mithun JM, Azad MAK, et al. Synthesis of Fe-or Ag-doped TiO_2–MWCNT nanocomposite thin films and their visible-light-induced catalysis of dye degradation and antibacterial activity. Research on Chemical Intermediates. 2018;**44**:2667-2683

[36] Lavorato C, Argurio P, Mastropietro TF, Pirri G, Poerio T, Molinari R. Pd/TiO2 doped faujasite photocatalysts for acetophenone transfer hydrogenation in a photocatalytic membrane reactor. Journal of Catalysis. 2017;**353**:152-161

[37] Jin C, Dai Y, Wei W, Ma X, Li M, Huang B. Effects of single metal atom (Pt, Pd, Rh and Ru) adsorption on the photocatalytic properties of anatase TiO_2. Applied Surface Science. 2017;**426**:639-646

[38] Zou Z, Zhou Z, Wang H, Yang Z. Effect of Au clustering on ferromagnetism in au doped TiO_2 films: Theory and experiments investigation. Journal of Physics and Chemistry of Solids. 2017;**100**:71-77

[39] Zhang J, Jin X, Morales-Guzman PI, Yu X, Liu H, Zhang H, et al. Engineering the absorption and field enhancement properties of Au–TiO_2 nanohybrids via whispering gallery mode resonances for photocatalytic water splitting. ACS Nano. 2016;**10**(4):4496-4503

[40] Sousa-Castillo A, Comesaña-Hermo M, Rodriguez-Gonzalez B, Pérez-Lorenzo M, Wang Z, Kong X-T, et al. Boosting hot electron-driven photocatalysis through anisotropic plasmonic nanoparticles with hot spots in Au–TiO_2 nanoarchitectures. The Journal of Physical Chemistry C. 2016;**120**(21):11690-11699

[41] Wang F, Wong RJ, Ho JH, Jiang Y, Amal R. Sensitization of Pt/TiO_2 using plasmonic Au nanoparticles for hydrogen evolution under visible-light irradiation. ACS applied materials & interfaces. 2017;**9**(36):30575-30582

[42] Hu Y, Song X, Jiang S, Wei C. Enhanced photocatalytic activity of Pt-doped TiO_2 for NOx oxidation both under UV and visible light irradiation: A synergistic effect of lattice Pt4+ and surface PtO. Chemical Engineering Journal. 2015;**274**:102-112

[43] Chowdhury P, Gomaa H, Ray AK. Sacrificial hydrogen generation from aqueous triethanolamine with eosin Y-sensitized Pt/TiO_2 photocatalyst in UV, visible and solar light irradiation. Chemosphere. 2015;**121**:54-61

[44] Huang B-S, Su E-C, Huang Y-Y, Tseng H-H. Tailored Pt/TiO_2 photocatalyst with controllable phase prepared via a modified sol–gel process for dye degradation. Journal of Nanoscience and Nanotechnology. 2018;**18**(3):2235-2240

[45] Liu Q, Yu ZB, Zhang RH, Li MJ, Chen Y, Wang L, et al. Photocatalytic degradation of perfluorooctanoic acid by Pd-TiO_2 photocatalyst. Huan jing ke xue= Huanjing kexue. 2015;**36**(6):2138-2146

[46] Repousi V, Petala A, Frontistis Z, Antonopoulou M, Konstantinou I, Kondarides DI, et al. Photocatalytic degradation of bisphenol a over Rh/TiO_2 suspensions in different water matrices. Catalysis Today. 2017;**284**:59-66

[47] Tahir M, Amin NS. Indium-doped TiO_2 nanoparticles for photocatalytic CO_2 reduction with H_2O vapors to CH_4. Applied Catalysis B: Environmental. 2015;**162**:98-109

[48] Inturi SNR, Boningari T, Suidan M, Smirniotis PG. Visible-light-induced photodegradation of gas phase acetonitrile using aerosol-made transition metal (V, Cr, Fe, co, Mn, Mo, Ni, cu, Y, Ce, and Zr) doped TiO_2. Applied Catalysis B: Environmental. 2014;**144**:333-342

[49] Sood S, Umar A, Mehta SK, Kansal SK. Highly effective Fe-doped TiO_2 nanoparticles photocatalysts for visible-light driven photocatalytic degradation of toxic organic compounds. Journal of Colloid and Interface Science. 2015;**450**:213-223

[50] Birben NC, Uyguner-Demirel CS, Kavurmaci SS, Gürkan YY, Turkten N, Cinar Z, et al. Application of Fe-doped TiO_2 specimens for the solar photocatalytic degradation of humic acid. Catalysis Today. 2017;**281**:78-84

[51] Yadav HM, Otari SV, Bohara RA, Mali SS, Pawar SH, Delekar SD. Synthesis and visible light photocatalytic antibacterial activity of nickel-doped TiO_2 nanoparticles against gram-positive and gram-negative bacteria. Journal of Photochemistry and Photobiology A: Chemistry. 2014;**294**:130-136

[52] Tripathi AK, Mathpal MC, Kumar P, Singh MK, Soler MAG, Agarwal A. Structural, optical and photoconductivity of Sn and Mn doped TiO_2 nanoparticles. Journal of Alloys and Compounds. 2015;**622**:37-47

[53] Khan H, Berk D. Synthesis, physicochemical properties and visible light photocatalytic studies of molybdenum, iron and vanadium doped titanium dioxide. Reaction Kinetics, Mechanisms and Catalysis. 2014;**111**(1):393-414

[54] Choudhury B, Choudhury A, Borah D. Interplay of dopants and defects in making cu doped TiO_2 nanoparticle a ferromagnetic semiconductor. Journal of Alloys and Compounds. 2015;**646**:692-698

[55] Kaviyarasu K, Geetha N, Kanimozhi K, Magdalane CM, Sivaranjani S, Ayeshamariam A, et al. In vitro cytotoxicity effect and antibacterial performance of human lung epithelial cells A549 activity of zinc oxide doped TiO_2 nanocrystals: Investigation of bio-medical application by chemical method. Materials Science and Engineering: C. 2017;**74**:325-333

[56] Ola O, Maroto-Valer MM. Transition metal oxide based TiO_2 nanoparticles for visible light induced CO_2 photoreduction. Applied Catalysis A: General. 2015;**502**:114-121

[57] Moradi V, Jun MBG, Blackburn A, Herring RA. Significant improvement in visible light photocatalytic activity of Fe doped TiO_2 using an acid treatment process. Applied Surface Science. 2018;**427**:791-799

[58] Bhethanabotla VC, Russell DR, Kuhn JN. Assessment of mechanisms for enhanced performance of Yb/Er/titania photocatalysts for organic degradation: Role of rare earth elements in the titania phase. Applied Catalysis B: Environmental. 2017;**202**:156-164

[59] Reszczyńska J, Grzyb T, Wei Z, Klein M, Kowalska E, Ohtani B, et al. Photocatalytic activity and luminescence properties of RE3+–TiO_2 nanocrystals prepared by sol–gel and hydrothermal methods. Applied Catalysis B: Environmental. 2016;**181**:825-837

[60] Mazierski P, Lisowski W, Grzyb T, Winiarski MJ, Klimczuk T, Mikołajczyk A, et al. Enhanced photocatalytic properties of lanthanide-TiO_2 nanotubes: An experimental and theoretical study. Applied Catalysis B: Environmental. 2017;**205**:376-385

[61] Ji L, Zhang Y, Miao S, Gong M, Liu X. In situ synthesis of carbon doped TiO_2 nanotubes with an enhanced photocatalytic performance under UV and visible light. Carbon. 2017;**125**:544-550

[62] Simsek EB. Solvothermal synthesized boron doped TiO_2 catalysts: Photocatalytic degradation of endocrine disrupting compounds and pharmaceuticals under visible light irradiation. Applied Catalysis B: Environmental. 2017;**200**:309-322

[63] Zhang L, Qingrui SUN, Hongxiao YAN. Method for Preparing Iodine-Doped TiO_2 Nano-Catalyst and Use Thereof in Heterogeneously Catalyzing Configuration Transformation of Trans-Carotenoids. Google Patents; 2017

[64] Yu W, Liu X, Pan L, Li J, Liu J, Zhang J, et al. Enhanced visible light photocatalytic degradation of methylene blue by F-doped TiO_2. Applied Surface Science. 2014;**319**:107-112

[65] Helmy ET, El Nemr A, Mousa M, Arafa E, Eldafrawy S. Photocatalytic degradation of organic dyes pollutants in the industrial textile wastewater by using synthesized TiO_2, C-doped TiO_2, S-doped TiO_2 and C, S co-doped TiO_2 nanoparticles. Journal of Water and Environmental Nanotechnology. 2018;**3**:116-127

[66] Jia T, Fu F, Yu D, Cao J, Sun G. Facile synthesis and characterization of N-doped TiO_2/C nanocomposites with enhanced visible-light photocatalytic performance. Applied Surface Science. 2018;**430**:438-447

[67] Lee HC, Park HS, Cho SK, Nam KM, Bard AJ. Direct photoelectrochemical characterization of photocatalytic H, N doped TiO_2 powder suspensions. Journal of Electroanalytical Chemistry. 2017;**819**:38-45

[68] Jaiswal R, Bharambe J, Patel N, Dashora A, Kothari DC, Miotello A. Copper and nitrogen co-doped TiO_2 photocatalyst with enhanced optical absorption and catalytic activity. Applied Catalysis B: Environmental. 2015;**168**:333-341

[69] Zeng L, Lu Z, Li M, Yang J, Song W, Zeng D, et al. A modular calcination method to prepare modified N-doped TiO_2 nanoparticle with high photocatalytic activity. Applied Catalysis B: Environmental. 2016;**183**:308-316

[70] Phongamwong T, Chareonpanich M, Limtrakul J. Role of chlorophyll in Spirulina on photocatalytic activity of CO_2 reduction under visible light over modified N-doped TiO_2 photocatalysts. Applied Catalysis B: Environmental. 2015;**168**:114-124

[71] Mahy JG, Cerfontaine V, Poelman D, Devred F, Gaigneaux EM, Heinrichs B, et al. Highly efficient low-temperature N-doped TiO_2 catalysts for visible light photocatalytic applications. Materials. 2018;**11**(4):584

[72] Wang F, He X, Sun L, Chen J, Wang X, Xu J, et al. Engineering an N-doped TiO 2@ N-doped C butterfly-like nanostructure with long-lived photo-generated carriers for efficient photocatalytic selective amine oxidation. Journal of Materials Chemistry A. 2018;**6**(5):2091-2099

[73] Samsudin EM, Hamid SBA, Juan JC, Basirun WJ, Centi G. Synergetic effects in novel hydrogenated F-doped TiO_2 photocatalysts. Applied Surface Science. 2016;**370**:380-393

[74] McManamon C, O'Connell J, Delaney P, Rasappa S, Holmes JD, Morris MA. A facile route to synthesis of S-doped TiO_2 nanoparticles for photocatalytic activity. Journal of Molecular Catalysis A: Chemical. 2015;**406**:51-57

[75] Lei XF, Xue XX, Yang H, Chen C, Li X, Niu MC, et al. Effect of calcination temperature on the structure and visible-light photocatalytic activities of (N, S and C) co-doped TiO_2 nano-materials. Applied Surface Science. 2015;**332**:172-180

[76] Noorimotlagh Z, Kazeminezhad I, Jaafarzadeh N, Ahmadi M, Ramezani Z, Silva Martinez S. The visible-light photodegradation of nonylphenol in the presence of carbon-doped TiO_2 with rutile/anatase ratio coated on GAC: Effect of parameters and degradation mechanism. Journal of Hazardous Materials. 2018;**350**:108-120

[77] Abdullah H, Khan MMR, Ong HR, Yaakob Z. Modified TiO 2 photocatalyst for CO 2 photocatalytic reduction: An overview. Journal of CO_2 Utilization. 2017;**22**:15-32

[78] Jin Q, Nie C, Shen Q, Xu Y, Nie Y. Cobalt and sulfur co-doped TiO_2 nanostructures with enhanced photo-response properties for photocatalyst. Functional Materials Letters. 2017;**10**(05):1750061

[79] Zhang Y, Han C, Zhang G, Dionysiou DD, Nadagouda MN. PEG-assisted synthesis of crystal TiO_2 nanowires with high specific surface area for enhanced photocatalytic degradation of atrazine. Chemical Engineering Journal. 2015;**268**:170-179

[80] Park J-Y, Kim C-S, Okuyama K, Lee H-M, Jang H-D, Lee S-E, et al. Copper and nitrogen doping on TiO_2 photoelectrodes and their functions in dye-sensitized solar cells. Journal of Power Sources. 2016;**306**:764-771

[81] Chatzitakis A, Grandcolas M, Xu K, Mei S, Yang J, Jensen IJT, et al. Assessing the photoelectrochemical properties of C, N, F codoped TiO_2 nanotubes of different lengths. Catalysis Today. 2017;**287**:161-168

[82] Zhao XG, Huang LQ. Iridium, carbon and nitrogen multiple-doped TiO_2 nanoparticles with enhanced photocatalytic activity. Ceramics International. 2017;**43**(5):3975-3980

[83] Tan D, Liu Y, Guo W, Wang J, Chang Z, Li D. Synthesis and characterisation of W–Bi–S-tridoped TiO_2 nanoparticles with enhanced photocatalytic activity. Materials Science and Technology. 2016;**32**(16):1673-1677

[84] Lin J, Liu X, Zhu S, Liu Y, Chen X. Anatase TiO(2) nanotube powder film with high crystallinity for enhanced photocatalytic performance. Nanoscale Research Letters. 2015; **10**:110

[85] Podporska-Carroll J, Panaitescu E, Quilty B, Wang L, Menon L, Pillai SC. Antimicrobial properties of highly efficient photocatalytic TiO_2 nanotubes. Applied Catalysis B: Environmental. 2015;**176-177**:70-75

[86] Weon S, Choi W. TiO_2 nanotubes with open channels as deactivation-resistant photocatalyst for the degradation of volatile organic compounds. Environmental Science & Technology. 2016;**50**(5):2556-2563

[87] Tang H, Zhang D, Tang G, Ji X, Li C, Yan X, et al. Low temperature synthesis and photocatalytic properties of mesoporous TiO_2 nanospheres. Journal of Alloys and Compounds. 2014;**591**:52-57

[88] Zhou X, Liu N, Schmuki P. Photocatalysis with TiO_2 nanotubes: "Colorful" reactivity and designing site-specific Photocatalytic centers into TiO_2 nanotubes. ACS Catalysis. 2017;**7**(5):3210-3235

[89] Tan L-L, Ong W-J, Chai S-P, Goh BT, Mohamed AR. Visible-light-active oxygen-rich TiO_2 decorated 2D graphene oxide with enhanced photocatalytic activity toward carbon dioxide reduction. Applied Catalysis B: Environmental. 2015;**179**:160-170

[90] Lin L-Y, Nie Y, Kavadiya S, Soundappan T, Biswas P. N-doped reduced graphene oxide promoted nano TiO_2 as a bifunctional adsorbent/photocatalyst for CO_2 photoreduction: Effect of N species. Chemical Engineering Journal. 2017;**316**:449-460

[91] Qu A, Xie H, Xu X, Zhang Y, Wen S, Cui Y. High quantum yield graphene quantum dots decorated TiO_2 nanotubes for enhancing photocatalytic activity. Applied Surface Science. 2016;**375**:230-241

[92] Tian H, Shen K, Hu X, Qiao L, Zheng WN. S co-doped graphene quantum dots-graphene-TiO_2 nanotubes composite with enhanced photocatalytic activity. Journal of Alloys and Compounds. 2017;**691**:369-377

[93] Mishra A, Mehta A, Sharma M, Basu S. Impact of ag nanoparticles on photomineralization of chlorobenzene by TiO_2/bentonite nanocomposite. Journal of Environmental Chemical Engineering. 2017;**5**(1):644-651

[94] Seftel EM, Niarchos M, Mitropoulos C, Mertens M, Vansant EF, Cool P. Photocatalytic removal of phenol and methylene-blue in aqueous media using TiO_2@ LDH clay nanocomposites. Catalysis Today. 2015;**252**:120-127

[95] He Y, Sutton NB, Rijnaarts HHH, Langenhoff AAM. Degradation of pharmaceuticals in wastewater using immobilized TiO_2 photocatalysis under simulated solar irradiation. Applied Catalysis B: Environmental. 2016;**182**:132-141

[96] Feng D, Xu S, Liu G. Application of immobilized TiO_2 photocatalysis to improve the inactivation of Heterosigma akashiwo in ballast water by intense pulsed light. Chemosphere. 2015;**125**:102-107

[97] Belver C, Han C, Rodriguez JJ, Dionysiou DD. Innovative W-doped titanium dioxide anchored on clay for photocatalytic removal of atrazine. Catalysis Today. 2017;**280**:21-28

[98] Mishra A, Mehta A, Sharma M, Basu S. Enhanced heterogeneous photodegradation of VOC and dye using microwave synthesized TiO_2/clay nanocomposites: A comparison study of different type of clays. Journal of Alloys and Compounds. 2017;**694**:574-580

Chapter 4

Modified Metallic Oxides for Efficient Photocatalysis

Vladimir A. Escobar Barrios,
Dalia Verónica Sánchez Rodríguez,
Nancy Ayerim Cervantes Rincón and
Alma Berenice Jasso-Salcedo

Additional information is available at the end of the chapter

http://dx.doi.org/10.5772/intechopen.80834

Abstract

The aim of the chapter is to present modified materials like alternatives for conventional photocatalyst such as titanium dioxide. Discussion about silver/graphene nanoparticles-modified zinc oxide for the degradation of pollutants like triclosan or bisphenol A, both considered as endocrine disruptors, which affect the hormonal development of humans, is presented. The best conditions to obtain the highest photodegradation degree are established. In addition, the bismuth oxychloride has gained attention during the last 5 years for photocatalysis. In accordance, the obtained results for phenol photodegradation, using such oxychloride, are also presented. In the chapter, the characterization of photocatalyst is reported along with the proposal for mechanisms of action for the modified ZnO photocatalyst and the bismuth oxychloride.

Keywords: ZnO, bismuth oxychloride, silver and graphene doped

1. Introduction

The photocatalysis has become an important research area in the recent years. The interest on such area has accelerated the number of papers, since photocatalysts have diverse applications, for example, the photodegradation of several contaminants that are quite difficult to transform to innocuous substance as water vapor. Other important applications are the development of self-cleaning products as well as the production of highly effective fuels like hydrogen. Nowadays, the technological development based on photocatalysis has an important growth in regions such as Europe and Asia (12% annually), with an estimated market of about € 1500 million in 2015 [1].

Photocatalysts are based on semiconductor materials, which are activated by radiation with higher energy than the bandgap of the used semiconductor, in order to create hole-electron pairs, once the electron passes from valence to conduction band. The promoted electron, toward conduction band, and the hole remained in valence band react in the photocatalyst interface with adsorbed substances in order to create reactive entities (free radicals and/or radical anions), which interact with contaminants to degrade them. All these processes imply the sorption on the photocatalyst's interface, its activation by radiation, creation of reactive species, and the recombination of electron to hole. The last could occur very fast that the creation of reactive species does not take place at great extent, giving ineffective photodegradation process.

In this context, the research regarding photocatalysis, in recent years, is focused to decrease the bandgap of photocatalyst in order to use solar light rather than UV light, which implies additional cost during the treatment process. In addition, minimization of the recombination process, in order to improve the photocatalytic performance, is quite important too. Some approaches to face these issues consider the doping with metals; meanwhile other research groups also consider the modifications with electron-rich species like graphene.

TiO_2 has been the photocatalyst by excellence and has been widely studied in its pristine form and/or doped with metals like gold and other elements. Nevertheless, other metal oxides, such as ZnO, have shown better photocatalytic performance, especially when visible light is used. ZnO has also been doped with metals like gold or silver. More recently, other metal oxides like bismuth oxychloride have gained attention for photocatalytic process.

In this chapter, discussion about the modification of ZnO with graphene is presented and discussed in terms of the implied mechanism (hybridization), in addition to the obtained results when such hybrid photocatalyst was used for the photodegradation of triclosan under visible light. Additionally, the modification method of ZnO with silver nanoparticles and its effect on photocatalytic performance for bisphenol A, Rhodamine B (RhB), and Triclosan is presented.

Finally, the use of an attractive photocatalyst, bismuth oxychloride (BiOCl), and its respective modification with silver and graphene oxide for RhB photodegradation along with the result and mechanism for the photocatalyst based on TiO_2-BiOCl used for photodegradation of phenol is presented and discussed.

The three different photocatalysts show how effective photocatalyst can be obtained and modified. The presented and discussed results contribute to understand some of the key parameters implied in the photocatalysis.

2. Zinc oxide

Zinc oxide (ZnO) is a II-VI group semiconductor with wurtzite structure with lattice parameters a = 0.3296 nm and c = 0.52065 nm. Zn^{2+} atoms are tetrahedrally coordinated with O^{2-} atoms stacked alternately along the *c* axis so that d-electrons of zinc are hybridized with 2p-electrons of oxygen [2]. ZnO has the potential to become an important material for photocatalysis because it is nontoxic (it is often included in pharmaceuticals), photostable, and low cost.

We consider at least three windows of opportunity to boost ZnO as photocatalyst in academy and industry. The first is related to the quantum yield in visible light rather than UV for the efficient utilization of the electron-hole (e^-/h^+) pairs during decomposition of endocrine disruptors and emergent contaminants in water. A second opportunity is the immobilization of the photocatalyst that could make possible its utilization on continuous mode and toward implementation at industrial scale. The third challenge deals with nonlinear mathematical modeling of the factors that rule the photocatalytic kinetics of the metallic-modified ZnO for optimization purposes. This is not a trivial problem in photocatalysis and artificial neural network tools (ANNs) do not demand of academic expertise for a fast implementation; for example, in the industry sector. The reader can consult [3] for the approach that Meimaroglou and collaborators follow to assess ZnO photocatalyst structure-photocatalytic performance associations toward the maximization of the photocatalytic efficiency. The first two challenges are addressed in the present chapter.

2.1. Use of ZnO/graphene photocatalysts

Nowadays, solar energy, radiant light, and heat from the sun are the most abundant available sources of clean energy. Thus, research studies and development of materials that can efficiently harvest solar irradiation and used for green environmental pollution management are essential. Photocatalysis, which could use renewable solar energy to activate the chemical reactions via oxidation and reduction, such as that occurs in advance oxidation processes (AOPs), is a sustainable technology to provide solution for environmental issue. This photocatalysis system has attracted great interest from science community as the most promising way to solve the environmental problems, especially getting rid of residual pollutants from wastewater stream.

In the field of photocatalysis, ZnO has emerged as the leading candidate for green environmental management systems because of its unique characteristics, such as wide bandgap (3.37 eV) in the near-UV spectral region, a large electron exciton binding energy of 60 eV at room temperature, strong oxidation ability, and good photocatalytic property [4]. It is a well-known fact that ZnO occurs as white hexagonal crystal or white powder known as white zinc. ZnO crystallizes in the wurtzite structure and is available as large bulk single crystals [5]. As an important semiconductor material, ZnO has been applied in catalysis, rubber and paint industries, ceramic bodies, varistors, fertilizers, and cosmetics [6].

Recently, the development of ZnO with precisely controllable features has gained significant scientific interest. The electrical, optical, and magnetic properties of ZnO can be altered or improved by the use of ZnO in nanoscale and efforts have been developed to improve the properties of ZnO photocatalyst [7]. Moreover, ZnO is an environmental friendly material as it is compatible with living organisms [8]. Since ZnO has almost the same bandgap energy as TiO_2 (3.2 eV), its photocatalytic capability is anticipated to be similar to that of TiO_2. However, ZnO is relatively cheaper compared to TiO_2, whereby the usage of titanium dioxide is uneconomic for large-scale water treatment operations [9]. The greatest advantage of ZnO is the ability to absorb a wide range of solar spectrum and more light quanta than some semiconducting metal oxides including the capacity to absorb visible light energy, which is due to its wide band energy. This results in fast recombination of photogenerated charges and thus caused low photocatalytic efficiency.

The recombination of photogenerated hole (hVB^+) and electron (eCB^-) is one of the major disadvantages in semiconductor photocatalysis. This recombination step lowers the quantum yield and causes energy wasting. Therefore, the e^-/h^+ recombination process should be inhibited to ensure efficient photocatalysis. Metal doping could counter the recombination problem with efficient charge separation between electrons and holes in ZnO photocatalyst. In addition, the dopants may trap electrons, reducing the chances of e^-/h^+ recombination that deactivate the photocatalytic system [10]. Furthermore, the generation of hydroxyl radicals and active oxygen species will greatly increase resulting from the enhancement in charge separation efficiency [11]. Semiconductors as graphene have been proven as a couple semiconductor that can improve the visible-light photocatalytic efficiency of ZnO due to its remarkable chemical, physical, and mechanical properties, such as large surface area (2600 m^2/g), excellent electrical and thermal conductivity, high mechanical strength, flexibility, and efficient wide range of light adsorption. Due to the properties of graphene-based materials, several ranges of environmental applications have been developed such as absorption, transformation, and detection [12]. So far, numerous methods have been used to design and synthesize ZnO/graphene hybrid photocatalysts with various morphologies. However, most of these methods rely on chemical and/or high-energy consumption resulting in a costly, environmentally hazardous, and especially inefficient photocatalyst for complete degradation of organic pollutants as triclosan, which has been classified as potential endocrine disrupting compound (EDC). Triclosan was ranked as the most abundant compound among all investigated pharmaceuticals and personal care products with its mean concentration of 12.6 ± 3.8 mg/Kg in 110 biosolids samples collected from 94 US wastewater treatment plants across 32 states and the District of Columbia using EPA Method 1694 [13]. In addition, the highest initial concentrations of triclosan detected in municipal biosolids were 2715 and 1265 μg/Kg [14].

Hydrothermal and chemical reductions have been the main methods studied to obtain ZnO/graphene hybrid photocatalysts. However, photodeposition method [15] has been reported as an efficient method to generate hybrid photocatalyst to degrade pollutants. This study proposes the synthesis of ZnO/graphene semiconductors by developing a facile, cheap, environmentally and high reproducibility approach to obtain an efficient material to degrade organic pollutants, as triclosan (TCS), under visible light. For instance, it has been demonstrated that inductive irradiation method is possible to synthesize ZnO composites due to its polarity [16]. In particular, graphene has the ability to accept electrons efficiently due to the absence of oxygen chemical groups on its surface preventing the recombination and providing a favorable π-π conjugation between TCS and aromatic region of graphene (**Figure 1**). The trapped electrons on graphene react with the dissolved oxygen and water to form reactive superoxide and hydroxyl radicals, which further oxidizes triclosan.

The modified photodeposition method has resulted in a successful process to prepare ZnO/graphene hybrid photocatalysts with enhanced photocatalytic activity under visible light radiation. The obtained results show degradation of 1% of triclosan (8 ppm) in absence of catalyst (photolysis) while this degradation percentage increases up to 42% using 0.5 wt% of

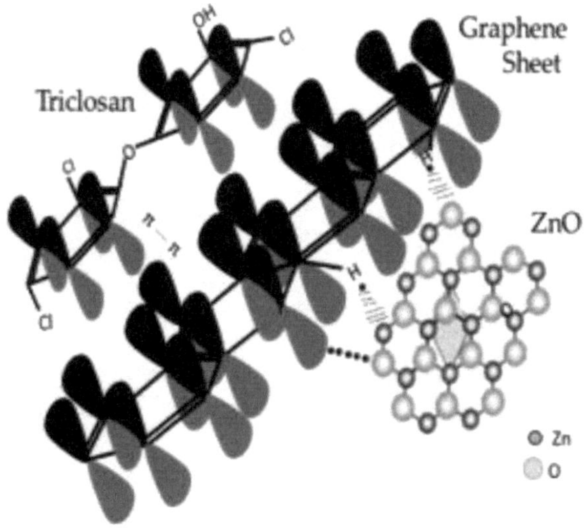

Figure 1. Schematic illustration of the interaction between ZnO photocatalyst doped with graphene sheets and its interaction with triclosan.

graphene sheets as dopant in ZnO nanocatalyst. It is important to mention that wurtzite has been the structure used in this study with a bandgap of 3.21 eV obtained by UV-vis spectrophotometer, while a value of 3.15 eV for ZnO/graphene composite was obtained. This reduction in the bandgap value is associated to a structure based on good interaction between ZnO nanocatalyst and graphene sheets, which creates intermediate energy levels between both materials, a property that allows to transfer the electrons promoted from the valence band to the conductive band of the ZnO semiconductor to graphene sheets that capture and retain the transferred electrons, improving the photocatalytic performance of ZnO as it is illustrated in **Figure 2**. This important first approximation has been the result of the evaluation of the effect of the loaded graphene amount in zinc oxide nano-photocatalyst. Graphene concentrations of 0.25 and 0.5 wt% were evaluated for this purpose and the obtained results showed enhancement of the ZnO photocatalytic activity under visible light radiation even using minimum contents of graphene sheets. Thus, 34 and 36% of the initial concentration of triclosan (8 ppm) was degraded using ZnO/graphene composites loaded with 0.25 and 0.5 wt% of graphene, respectively, and the bandgap values were 3.19 and 3.18 eV, for such ZnO/graphene composites. In contrast with 25% of initial triclosan degraded using pristine ZnO (bandgap: 3.21 eV). These results can be compared with those reported in previous investigations related with the photodegradation of triclosan using dopants such as rare-earth elements as Ce (47%) [17], metals such as Au (10% after 5 h) [18], Ag [19], and Cu [20]; as well as oxide compounds as MgO, WO_3, TiO_2, ZnO, or graphene oxide [21] using dopant contents up to 10%.

The adsorption properties of the as-prepare ZnO/graphene hybrid photocatalysts doped with different amounts of graphene sheets were one of the most important characteristics to improve the degradation efficiency of the ZnO. The specific area of ZnO and ZnO/graphene hybrid photocatalysts was determined via N_2 adsorption isotherms using the Brunauer Emmett Teller (BET) method. The results revealed that the graphene monolayers (442 g/m^2) showed the highest surface area for the analyzed pristine material, while ZnO had a specific area of 10.8 g/m^2. Among the three investigated composites, the photocatalyst loaded with 0.5 wt% of graphene had the highest surface area (18.3 g/m^2), followed by ZnO/graphene 0.25 wt% (14 g/m^2) and ZnO/graphene 0.1 wt% (13.3 g/m^2) catalysts. Thus, the addition of graphene sheets increases the surface area of the hybrid catalysts up to 69% of the initial area of zinc oxide and improve their adsorption capacities, resulting in the first property of the as-prepare hybrid composites to increase the degradation of triclosan, as it has been reported previously [22].

In order to study the effect of the method of synthesis in the ZnO/graphene hybrids, composites were prepared by impregnation method with graphene contents of 0.5 wt% using continuous stirring in the absence of UV radiation. The resultant photocatalysts were tested to degrade triclosan under visible light radiation, and it was obtained that 15% of the initial concentration of triclosan was degraded, 53% less than the composites synthesized by photodeposition method as it can be observed in **Figure 2**.

The hybridization of ZnO with graphene has been studied and confirmed by Raman spectroscopy integrating the intensity ratio of the D to G bands characteristic for carbon materials. The G peak arises from the stretching of C-C bond of graphite materials and is highly sensitive to strain effects in sp^2 system, while D peak is caused by the disordered structure of graphene material. Regarding the sp^3 and sp^2 hybridizations I_D/I_G value, in the case of pristine graphene oxide was 0.93, indicating that the intensity of the G band is higher than D band, which results in a lower amount of sp^3 defects and less structural disorder. Compared with graphene oxide (0.93), the reduction of I_D/I_G ratio for ZnO/GO 0.5 wt% hybrid photocatalyst (I_D/I_G = 0.91) is observed, implying a reduction of sp^3 defects compared with pure graphene oxide. This fact

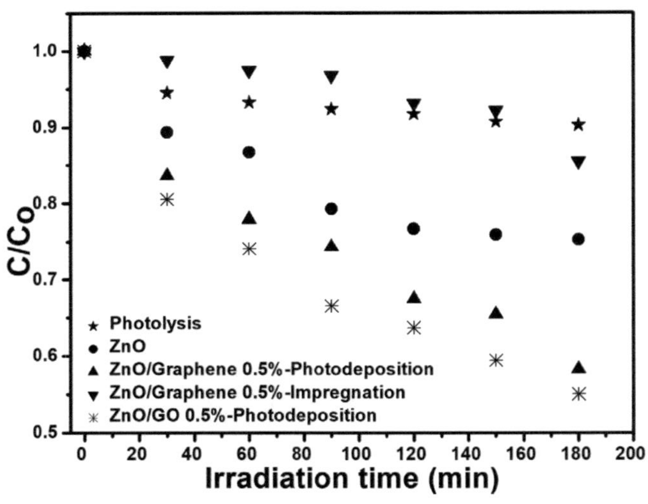

Figure 2. Degradation curves of triclosan under visible light radiation conditions: 23°C, pH = 7.

confirms the interaction between ZnO and GO given by sp^3 defects. The results given herein-above are consistent with the results in FTIR characterization, revealing the reestablishment of the conjugated graphene oxide network. In contrast, the ratio I_D/I_G for the composite prepared with graphene sheets was calculated and corresponds to the value of 1.04 (higher than I_D/I_G of pristine graphene, 0.36) indicating the presence of more defects in the graphene oxide lattice, which implies a decrease in the size of the in-plane sp^2 domains and formation of the defects and disorders in the graphene sheets, revealing the reestablishment of the conjugated graphene network (sp^2 carbon) [23] in the ZnO/GO photocatalysts due to the hybridization of graphene via photoirradiation method but resulting in a weak interaction due to the absence of chemical groups on the surface of graphene sheets as are presented on GO surface. The interaction between graphene oxide sheets was improved by the interaction of ZnO-polarized structure and the chemical surface of graphene oxide, which contains carbonyl, carboxyl, epoxy, and hydroxyl groups.

The degradation curves of triclosan for ZnO/GO hybrid photocatalyst are presented in **Figure 3**. It is noticed that the composites obtained by the photodeposition method are the materials with the best performance to degrade triclosan under visible light radiation compared with photolysis experiment and with the degradation using the bare ZnO photocatalyst.

In addition to the photodegradation curves presented in **Figure 3**, the rate constants are presented in **Table 1**.

The results show the highest apparent degradation rate constant for the hybrid catalyst synthesized with graphene oxide (0.003 min^{-1}), almost four times higher than the pristine ZnO semiconductor, corroborating the importance of interaction between ZnO and GO.

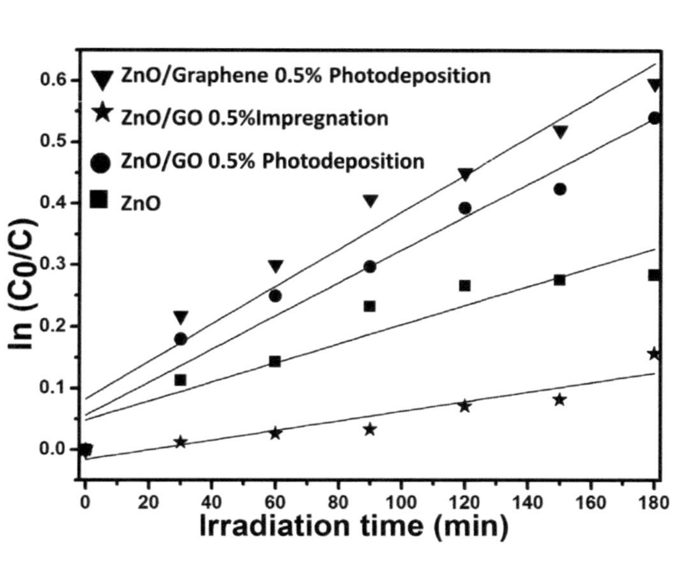

Figure 3. Triclosan photodegradation curves with ZnO/graphene and ZnO/GO hybrid photocatalysts under visible light. Conditions: 23°C, pH = 7.

Photocatalyst	K_{app}, min^{-1}
ZnO	0.00078
ZnO/G $_{PD}$	0.0016
ZnO/ OG $_{IMP}$	0.0027
ZnO/ OG $_{PD}$	0.003

Table 1. Apparent rate constants for triclosan photodegradation under visible light.

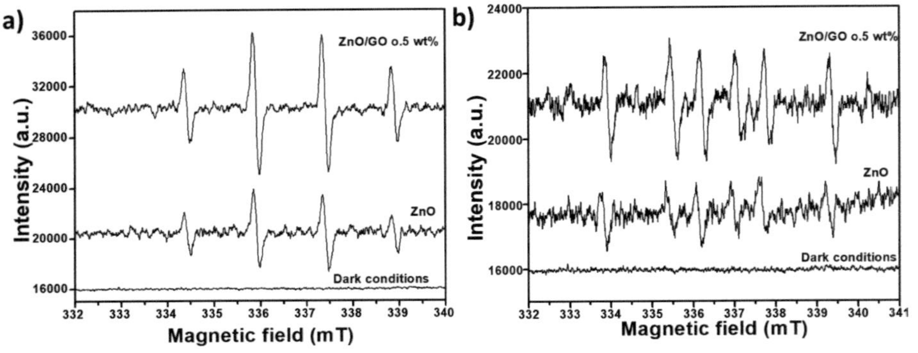

Figure 4. EPR spectra of ZnO/GO hybrid photocatalyst irradiated under visible light in (a) water and (b) ethanol solvents. DMPO was used as a radical trapper.

The mechanism of the best as-prepare ZnO/GO hybrid composite was proposed. Thus, the ESR spin-trap technique (with DMPO) was used to monitor the reactive oxygen species generated during the irradiation of the hybrid photocatalyst and the results are shown in **Figure 4**. Both signals of DMPO-$^{\bullet}$OH and DMPO-$O_2^{\bullet-}$ are clearly observed when pristine ZnO and ZnO/GO were exposed to visible radiation. Therefore, a dual mechanism involving both hydroxyl radicals and superoxide radicals is expected in the photocatalytic process. However, the signals for hybrid photocatalysts are stronger than the signals in pristine ZnO, thus accounting for the higher and stable photocatalytic performance of hybrid composites than bare ZnO toward the degradation of triclosan. The hydroxyl radicals trapped by DMPO (DMPO-$^{\bullet}$OH) and superoxide radicals (DMPO-$O_2^{\bullet-}$) for ZnO and hybrid photocatalysts were characterized by detecting four characteristic signals in water, and six signals in ethanol for DMPO-$^{\bullet}$OH and DMPO-$O_2^{\bullet-}$, respectively. In both cases, it is noticed that the signals for hybrid material are more pronounced than those for the ZnO pristine sample, thus accounting for the better photocatalytic performance.

Thus, the enhanced photocatalytic activity of ZnO/GO photocatalyst is due to the introduction of carbon material, which promotes an increase in charge separation to effective utilization of electrons to produce more $^{\bullet}$OH and $O_2^{\bullet-}$ radicals. In this case, the signals of $^{\bullet}$OH radicals are stronger than those for $O_2^{\bullet -}$, which suggests the predominance of oxidative reactions whose holes are responsible for the degradation of triclosan. Furthermore, ZnO photocatalyst presents visible light photocatalytic activity but generates stronger visible light after graphene hybridization, showing that graphene oxide is responsible for the visible light performance, which is induced by the injection of an excited electron from the lowest unoccupied molecular orbit (LUMO) of graphene to the conduction band (CB) of ZnO. The introduction of the graphene semiconductors can possibly cause the rapid separation of electron-hole pairs during irradiation [11] prolonging the electron-hole pair lifetime and accelerating the transfer rate of electrons [24] as shown in **Figure 5**.

In conclusion, the adsorption properties, good interaction between ZnO- and graphene-based materials, chemical structure of graphene, method of synthesis, and concentration of the dopant used to hybridize ZnO catalyst are the most important properties that affect the development of ZnO/graphene hybrid photocatalysts.

2.2. ZnO modification by inorganic molecules: silver

Silver nanoparticles are linked to the ZnO surface through the alkanethiol surfactant (**Figure 6a**). The stabilizer ligands COOH-$(CH_2)_n$-S-Ag keep nanoparticles as small as 7 nm in the solution and 15–26 nm in the Ag/ZnO photocatalyst, corroborated by XRD; although TEM analysis shows Ag nanoparticles of spherical shape and defined boundaries smallest as 3 nm over ZnO surface (**Figure 6b-d**). We studied the effect of the pH and time on the functionalization of ZnO nanoagglomerates by two methods denominated as photodeposition (PD) and impregnation (IMP) [15]. We propose to replace the ambiguous terms "doping" with functionalization when spoke of superficial ZnO modification. For instance, the sample 1%Ag/ZnO-PD11,1 synthesized by the PD method using 1 wt.% Ag at pH 11 and 1 h under UV light, a functionalization yield of 100% was corroborated by elemental analysis by inductively coupled plasma spectrometry ICP-OES and SEM-EDX. The UV irradiation produces free radicals ($^{\bullet}OH$, $^{-\bullet}O$) that degrade the ligand and release the silver nanoparticles onto the ZnO surface (**Figure 6c**). Byproducts of surfactant decomposition accumulate in ZnO are observed by IR. This functionalization of the

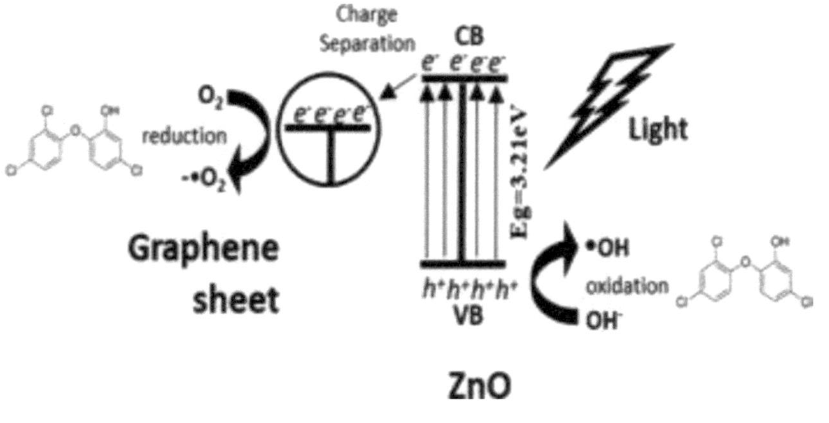

Figure 5. Schematic illustration of the photocatalytic process in the ZnO/GO hybrid semiconductor.

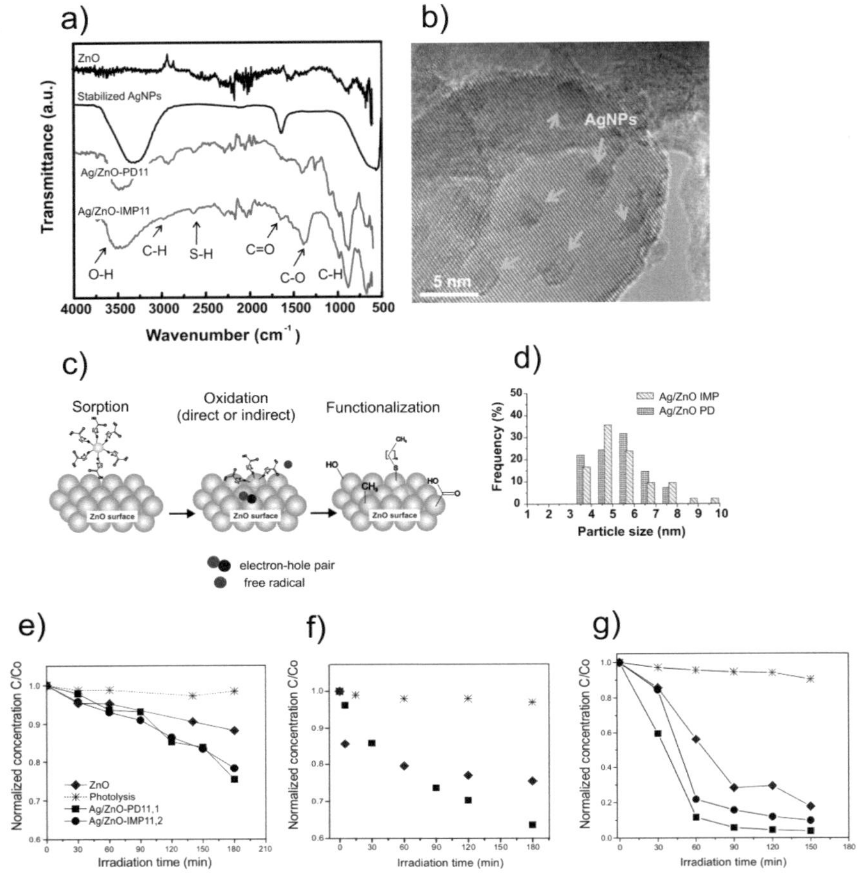

Figure 6. FTIR (a) and TEM image (b) of the as-synthesized photocatalyst functionalized ZnO by silver nanoparticles. Schematic illustration of the functionalization by PD method. (c). Histogram of silver particle size (d). Photocatalytic degradation of bisphenol A (e), triclosan (f), and RhB (g).

photocatalyst surface with S-H, C-O, and hydroxyl groups (^{-}OH) creates defect sites advantageous in photocatalysis. Silver-experienced redox processes during PD functionalization, for instance, oxidation (Ag^{+}/Ag^{0}, +0.799 eV vs. SHE) by photogenerated holes (+2.75 eV, vs. SHE) and free radicals. They reduce again into the zero-valence form (Ag^{0}) by photoexcited electrons. This method favors insertion of ionic silver Ag^{+} into ZnO crystalline structure perceived as an expansion of lattice parameters measured by XRD. The new attached silver possibly anchors on the surface defect sites of ZnO [25, 26]. The role of the metallic modifier in Ag/ZnO is to promote the pair electron-hole (e^{-}/h^{+}) separation and to increase the photocatalyst sensitivity toward visible light, in our case, evidenced as an absorption in the visible region by UV-Vis spectroscopy.

The IMP functionalization mechanism is different in the sense that silver nanoparticles and ZnO interaction is a function of the reaction time for an optimum of 2 h. For instance, the sample 1%Ag/ZnO-IMP11,2 synthesized using 1 wt.% Ag at pH 11 and 2 h under vigorous stirring results in heterogeneously distributed silver nanoparticles of average size of 15 nm determined by Scherrer equation. The surfactant decomposes almost completely during the heating (at 300°C) that releases AgNPs onto ZnO surface. As before, residuals S-H and C-H functionalizing the ZnO surface was observed by IR (**Figure 6a**). Similarly, the as-synthesized 1%Ag/ZnO-IMP samples absorb in visible region of the spectrum and silver nanoparticles induce a surface-localized plasmon resonance.

It is challenging to control the metallic particle size because the nanoparticles have the trend to form agglomerates. The deposition of silver nanoparticles over metallic oxides from preformed nanoparticles using, for example, chemical vapor deposition [27] has been used, but IMP and PD methods demonstrate to be a successful bulk functionalization of ZnO at room temperature and atmospheric pressure. The increase of photocatalytic efficiency by silver-modified ZnO is demonstrated on the degradation of endocrine disruptors (i.e., bisphenol-A), an emergent contaminant (i.e., triclosan), and a dye (i.e., RhB) under visible light, which represents an important achievement in the use of solar-driven Ag/ZnO photocatalysts.

Figure 6e-g shows the progress of the contaminant photodegradation under visible light (365 nm, 0.97 mW/cm^2, 8 W 3UV-38 UVP Inc. lamp, lab-made reactor) at 20°C. The results show that 25% of bisphenol-A (10 mg/L) was destroyed after 3 h using Ag/ZnO-PD, and it represents an improvement of 100% compared with ZnO. A total of 35% of initial triclosan (20 mg/L) was destroyed within 3 h using Ag/ZnO-IMP11, represents an improvement of 45% compared with ZnO. Finally, 90% of dye discoloration is obtained with Ag/ZnO-PD11, being only 20% better than ZnO.

Similarly, the photocatalysts were tested under UV light (302 nm). The use of BET surface area normalization of the apparent rate constant (k_{app}/BET ssa) in photocatalysis was proposed to clearly demonstrate that ZnO nanoagglomerates is 400% faster than Pi-25, a TiO_2 well-known and extensively used photocatalyst from Degussa. The bandgap energy of ZnO and titanium dioxide (TiO_2) is basically the same (3.2 eV). However, the valence and conduction bands exhibit differences in electric potential values; whose reported values are in the range of −0.45 to 2.75 eV vs. NHE and − 0.1 to 3.1 eV vs. NHE, for ZnO and TiO_2, respectively [28, 29]. Thus, the photogenerated holes in ZnO have strong enough oxidizing power to decompose most organic compounds. Furthermore, 1%Ag/ZnO-PD11,1 photocatalytic activity is 370% faster than ZnO attributed to the photoexcited electron trapping in the metallic primary (e^-/h^+ pair) and secondary active species (free radicals, i.e., $^{\bullet}OH$) for the oxidation of bisphenol-A. The effect of pH, photocatalyst dosage, and bisphenol-A concentration on the kinetic rate constant were also studied extensively and reported [3].

2.3. ZnO immobilization by organic molecules: poly (acrylic acid)

The utilization of powder photocatalyst may end unpractical for industrial scales because of the technical challenges like efficient dispersion and finally difficult separation of photocatalyst after the reaction that may entail important energetic costs and sometimes even producing a secondary pollution. Besides, photocorrosion is an important drawback in photocatalysis, and the anchoring of silver has been proved to control its progress. The immobilization of ZnO into organics tends to solve both the difficulty of photocatalyst dispersion and recuperation for a cyclic usage together with an improvement of the photostability [30]. The organics that have been reported for immobilization of TiO_2 comprise polyaniline [31, 32] and polypyrrole [33] that are hydrophobic and opaque, then it is incompatible for aqueous applications. Hydrogels, based on acrylic polymers, allow effective transport of water and other dissolved molecules due to their hydrophilicity and high swelling capacity. This kind of polymers shows stimuli-responsive properties to pH, temperature, solvent composition, and ionic strengths. Hydrogels come to be an ideal choice for Ag/ZnO photocatalyst immobilization since they are colorless and visually transparent, which permits penetration of light (**Figure 7e**).

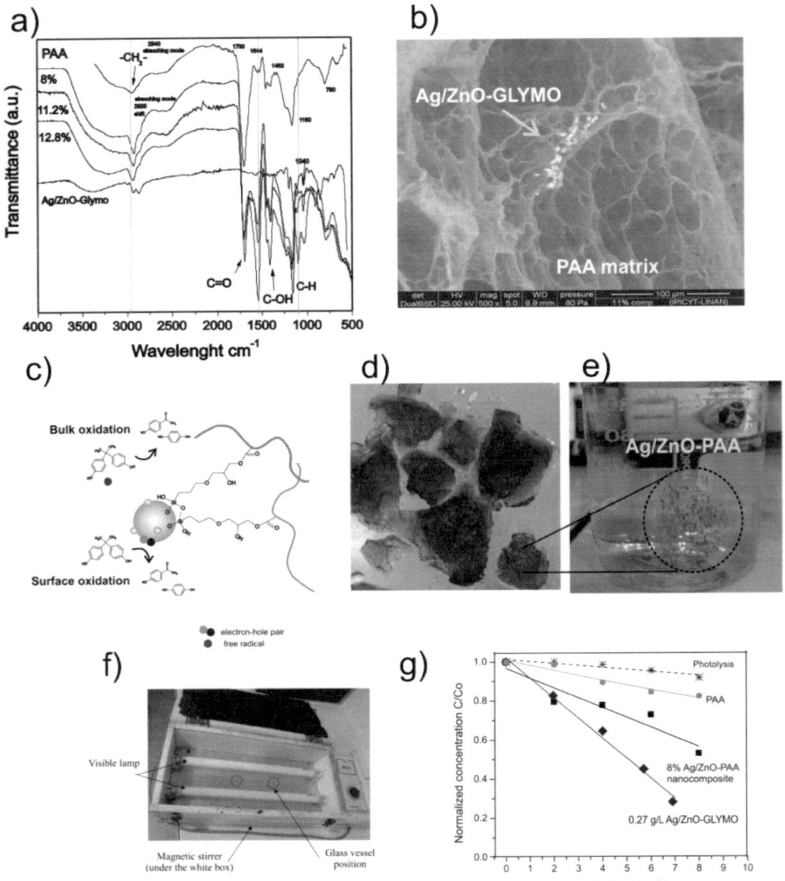

Figure 7. FTIR (a) and E-SEM image (b) of the as-synthesized Ag/ZnO-PAA. Schematic representation of the ligands between photocatalyst pending from the PAA chain and its photocatalytic role (c). Semi-swell Ag/ZnO-PAA composite (d) and reaction vessel containing photocatalyst dissolved in the pollutant solution (e).Home made photocatalytic reactor operated at room temperature and pressure (f). Photocatalytic degradation of bisphenol-A (g).

The general mechanism for the immobilization of Ag/ZnO into a poly (acrylic acid) (PAA) matrix involves the use of GLYMO as photocatalyst stabilizer and coupling agent. Infrared analysis (FTIR) shows characteristic bands at 916 and 1260 cm^{-1} for free epoxide functional groups, in pure GLYMO and Ag/ZnO-GLYMO, meaning that epoxide do not react with Ag/ZnO surface, as desired. Disappearance of bands at 823 and 780 cm^{-1}, corresponding to C-H from the methoxysilyl groups (CH_3O^-) of GLYMO, indicates two processes: hydrolysis of methoxy group and reaction with Ag/ZnO surface by weak bond such as hydrogen bonds. In addition, the peak at 1030 cm^{-1} corresponds to the formation of self-assembled monolayer Si-O-Si on Ag/ZnO surface, and confirms the efficient silanization of the photocatalyst (**Figure 7a**). Lastly, the interaction between the free tail of Ag/ZnO-GLYMO (highly reactive epoxy groups) and the carboxylic groups of PAA gives strong interaction and immobilization into the polymeric matrix.

Two possible arrangements of the photocatalyst bonded to the polymer matrix are proposed: (i) a photocatalyst pending from the PAA chain (**Figure 7c**) and (ii) a cross-linked–like structure, where the photocatalyst is the bridge between two PAA chains.

Environmental SEM chamber allow us for first time to obtain images of 3D network of cross-linked PAA under low vacuum that gives us a time-window of approximately 30 minutes for the analysis before dehydration (**Figure 7b**). SEM-EDX confirms the homogeneous dispersion of photocatalyst within the PAA matrix on 8%Ag/ZnO-PAA composites. The swelling capacity of the composite increases with photocatalyst content (5–13 wt.%) due to the high hydrophilicity of Ag/ZnO (**Figure 7d**). In addition, the composite photostability after 16 h of UV (365 nm) exposure was corroborated by FTIR and TGA analysis.

This enhanced photochemical stability has the potential of use as resistant composite packing material for continuous treatment of water under UV irradiation. This last aspect was first tested in batch experiments toward the degradation of bisphenol-A. Thus, **Figure 7g** shows the time evolution of the bisphenol-A photodegradation under visible light (>450 nm, 8 uW/cm^2, 8 W Hampton Bay lamp, home-made reactor in **Figure 7f**) at 50°C. On the other hand, 47% of the initial concentration of 10 mg/L is degraded within 7 h by 8%Ag/ZnO-PAA composite, it represents a decrease of 50% compared with Ag/ZnO-GLYMO. The composite was reused in a second consecutive cycle without washing that results in an improvement of 40% compared with the first cycle. It is necessary to highlight that the sorption of bisphenol-A in dark conditions was not important; therefore, regeneration of the composite is not necessary in continuous water treatments. Ag/ZnO-PAA composites like those synthesized in this study are less sensitive to saturation compared to zeolites and carbon materials.

3. Bismuth oxychloride (BiOCl)

Bismuth oxyhalides BiOCl (X:Cl, Br, and I) are a new class of semiconductors that have recently attracted attentions in the photocatalytic process due to their relatively slow electron-hole recombination process. BiOXs are conformed by Bi^{3+}, O^{2-}, and halide (X^-) ions stacked in $[X\text{-}Bi\text{-}O\text{-}Bi\text{-}X]_n$ layers, giving a tetragonal structure with no linkers interactions with halide

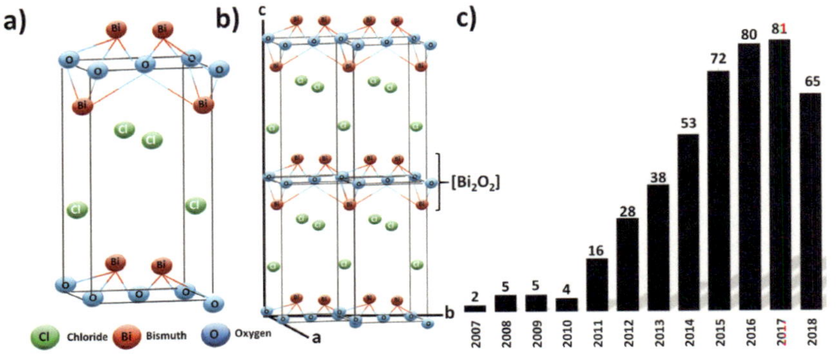

Figure 8. Scheme of unit cell (a), crystal structure BiOCl with {001} facet (b) and growing of articles of BiOCl in the photocatalytic process in later years (c).

along the C axis as showed in **Figure 8a, b**. The photocatalytic activity is due to [Bi_2O_2] layers intercalated of double halogen atoms that allow a better separation of the electron-hole pairs. The bismuth oxychloride (BiOCl) is a solid inorganic compound, not toxic, of pearlescent white color and due its brightness has been used in the cosmetic industry. There is a record not greater than 500 articles reported in the Scopus database concerning the use of BiOCl, as photocatalyst; this review was realized in June 2018 (**Figure 8c**) and the first publications appeared in 2007 with a growing interest during the last years.

The BiOCl has a bandgap of 3.2–3.5 eV and the valence band is constituted of O_{2p} and Cl_{3p}; while the conduction band has Bi_{6p} according to density functional theory (DFT), whereby when excited, the electrons from Cl atoms are displaced at bismuth orbital [34].

Most studies of BiOCl photocatalyst are focused on enhancing the photocatalytic activity for environmental remediation, especially for dyes degradation during water treatment. Due to the bandgap, its photocatalytic activity, under UV irradiation, has shown a similar or better photocatalytic activity than TiO_2 [35]. Some "model" dyes used to evaluate the photocatalytic activity with BiOCl are methylene blue, methyl orange [36], and Rh B [37]. In spite of the bandgap of BiOCl, there are reports wherein demonstrate an excellent degradation of RhB under visible light [38].

However, some challenges remain for BiOCl as most photocatalysts also face; one challenge is its activation under visible light due to the limited absorption of such radiation. A variety of strategies have been employed to get better light absorption and to decrease the charge carrier recombination, such as the heterojunction, doping impurity, and metallization of the surface. Modifications consider the presence of metal elements or compounds like Ag [39, 40], AgCl [41], Fe [42], and Bi [43], also with materials carbon-based as graphene [44], graphene oxide [45], and with other semiconductors as Co_3O_4 [46], BiOI [47], and TiO_2 [48–50]. However, most researches evaluated the photocatalytic activity for degradation of dyes, considering the RhB in many cases. Other pollutant photodegradation studies are bisphenol A using Fe-BiOCl under visible light [42], sulfanilamide with BiOCl-RGO [51], and phenol under visible light with BiOCl-TiO_2 composite [50]. Nowadays, the BiOCl is being investigated to reduce of carbon dioxide (CO_2) [52].

3.1. Experimental section for BiOCl

The photocatalytic activity was evaluated using two commercial BiOCl samples, which were modified with silver (Ag), graphene oxide (OG), and TiO_2. The BiOCl samples were named as P2600 and SB. The P2600 sample is a hydrophobic sample while SB sample is hydrophilic sample.

3.1.1. Modification of BiOCl samples

The modification of BiOCl with Ag and OG was carried out by photodeposition method, which consisted in disperse 500 mg of BiOCl (P2600 and SB) in deionized water during 30 minutes, then the particles of the Ag or OG were added in different percentages; in the case of Ag were 2.0, 0.5, and 0.1% w/w and for OG were 0.5 and 0.1% w/w. Then, the sample was irradiated using a Q200 reactor (λ = 250 nm, 304 mW/cm^2) for 3 h with continuous stirring. Later, the samples were centrifuged and thoroughly water washed for several cycles, and finally, they were dried at 70°C for 24 h. The composites with TiO_2 were prepared by solvothermal method with different weight percentages of TiO_2 (25, 50, and 75% w/w).

3.1.2. Characterization

The crystal structure of the synthesized photocatalysts was analyzed by X-ray diffraction (XRD) using a DX8 advance diffractometer (Bruker) with: Cu Kα radiation, 35 kV, 25 mA, λ = 0.15418 nm over the 2θ range of 10–80° in a step of 0.02°s^{-1}. The analysis of surface morphology and microstructure of the samples was carried out by a QUANTA 200 environmental scanning electron microscope.

3.1.3. Photocatalytic activity

The photocatalytic activity was evaluated with two pollutants. The samples modified with Ag and OG were evaluated for RhB degradation under visible light, the RhB degradation was followed by UV-Vis spectrophotometer at 552 nm. The composites with TiO_2 were evaluated for Phenol degradation under visible light (Xenon lamp; Oriel 300 W; λ = 450 nm). Monitoring of phenol degradation was carried out by high performance liquid chromatography (HPLC) using a C18 column with a mobile phase of acetonitrile-water (27–75%) with a flow of 0.5 mL min^{-1}.

3.2. Results and discussion

Both pure BiOCl samples (P2600 and SB) showed overlapping flakes forming flower-like morphology, the size of flakes were of 10–60 microns approximately in P2600 samples (**Figure 9a**), while SB samples displayed a size of flakes about 9–34 microns (**Figure 9b**).

In the Ag-modified samples, small BiOCl-impregnated particles on the flakes were observed, in the case of the modified SB samples, a major impregnation of Ag particles was noticed (**Figure 9c**) in comparison with Ag-P2600 sample (**Figure 9d**); probably due to hydrophobic character of P2600 sample avoiding that Ag particles could have contact with the photocatalyst. Regarding OG-BiOCl samples, in both modified photocatalysts, it was observed

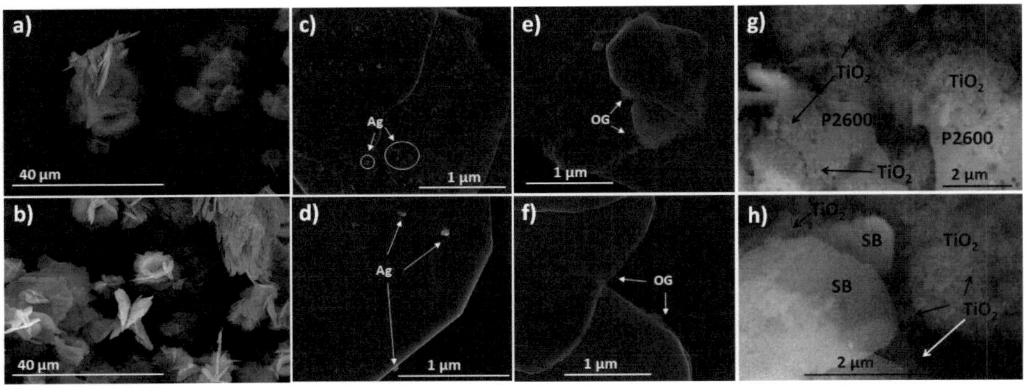

Figure 9. SEM images of BiOCl pure samples; SB (b) and P2600 (a). BiOCl modified with silver; Ag-SB (c) and Ag-P2600 (d). BiOCl modified with graphene oxide; OG-SB (e) and OG-P2600 (f). BiOCl modified with titanium dioxide; TiO_2-SB (g) and TiO_2-P2600 (h).

that small sheet of OG was deposited in the flakes of BiOCl, mainly in the edges of flakes as it is observed in **Figure 9e-f**. While in the composites of TiO_2-BiOCl, the generated TiO_2 was incrusted as small agglomerated particles on the BiOCl flakes (**Figure 9c-d**). Both photocatalysts could promote a better separation of photocharges generated during the reaction and therefore a better photocatalytic activity (**Figure 9g-h**).

The RhB degradation under visible light with Ag-BiOCl is shown in **Figure 10a**, **b**, the maximum degradation of RhB was obtained with pure BiOCl; and 30 and 40% of RhB degradation with P2600 and SB were obtained, respectively. In the case of Ag-SB, as the percentage of Ag increased, the photocatalytic activity decreased, while with Ag-P2600 was observed a greater degradation with 0.5% Ag than with 0.1%Ag. The presence of Ag in the BiOCl generates a decrease of photocatalytic activity, this result may be due to the presence of Ag in two oxidation states (Ag^0 and Ag^{1+}) observed in the XRD specter (**Figure 10c**, **d**), which may be acting as recombination sites. The modification with Ag incites a displacement of the [001] peak in the XRD specter probably due to the change of atoms of Bi by atoms of Ag because both elements have a similar ionic radius. Likewise, change of intensity in the [002] and [101] peaks in the spectra may be associated with a modification of the crystalline phase. This change in the Ag-BiOCl may affect the photocatalytic activity.

With the OG-BiOCl sample, the photocatalytic activity was different, since with low percentage of OG, the photocatalytic activity increased. The OG-SB sample (0.1% of OG) gave the greater degradation of RhB, while the OG-P2600 (0.1% of OG) and pure P2600 had the same degradation percentage (**Figure 11a**, **b**). In the XRD spectra of OG-BiOCl was observed, a decrement in intensities of [001], [102], and [112] peaks, and an intensity increment for [002] peak as the OG percentage increased. This result indicates that OG induces a better orientation of {001} facet in both BiOCls (**Figure 11c**, **d**).

Referring to TiO_2-BiOCl composite, its photocatalytic activity was evaluated for phenol degradation under visible light. The better composites were TiO_2-P2600 (75–25%) giving 45% of

http://dx.doi.org/10.5772/intechopen.80834

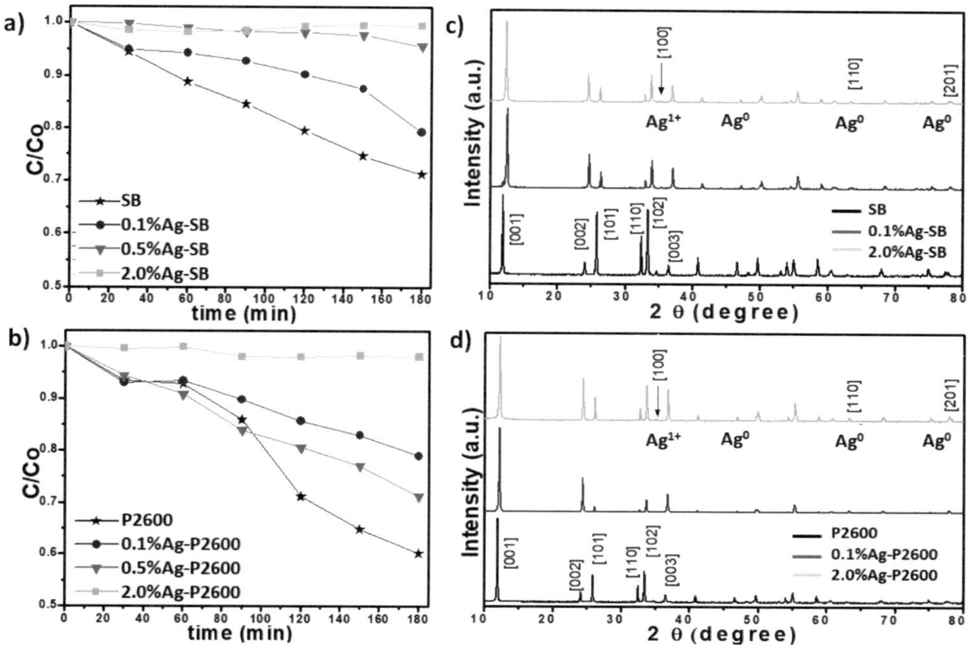

Figure 10. RhB degradation under visible irradiation with BiOCl pure and modified with silver; Ag-SB (a) and Ag-P2600 (b). XRD spectra of BiOCl samples; Ag-SB (c) and Ag-P2600 (d).

phenol degradation and TiO_2-SB (50–50%) giving 36% of phenol degradation during 6 h of reaction (**Figure 12a**, **b**). For the composite TiO_2-P2600 conforming the percentage of TiO_2 decreased, the photocatalytic activity also decreased. Such results confirm that there is a good interaction between both photocatalysts (BiOCl and TiO_2), in the TiO_2-P2600 composites, the size and hydrophobicity of P2600 played an important role in the adsorption of phenol on the surface, and therefore, gave a higher photocatalytic activity. In addition, higher TiO_2 percentage generated the appropriate heterojunction. In the XRD spectrum of TiO_2-BiOCl, the presence of peaks from both photocatalysts was observed (**Figure 12c**, **d**), also an overlap of the peaks [002] and [101] of BiOCl with the peak of [001] of TiO_2. In TiO_2-SB composites, it observed a lower intensity for [001] peak corresponding to BiOCl. Regarding TiO_2-SB composites, the SB exhibited a smaller size and hydrophilic character; then it was necessary that less amount of TiO_2 is active under visible light, and an increment of TiO_2 could generate the recombination of electron-hole pairs decreasing the photocatalytic activity.

The results obtained in our study offer a promising direction for the design of more practical and efficient photocatalysts to be used under visible light. In addition, the photocatalysis using BiOCl is becoming a promising research topic due to its fascinating characteristics, and it is necessary to study and understand the synthesis methods, morphology, predominant facets that improve the photocatalytic activity in all visible irradiations in order to achieve the mineralization of pollutants. Currently, our research group is working with the generation of biofuels by photocatalysis using BiOCl.

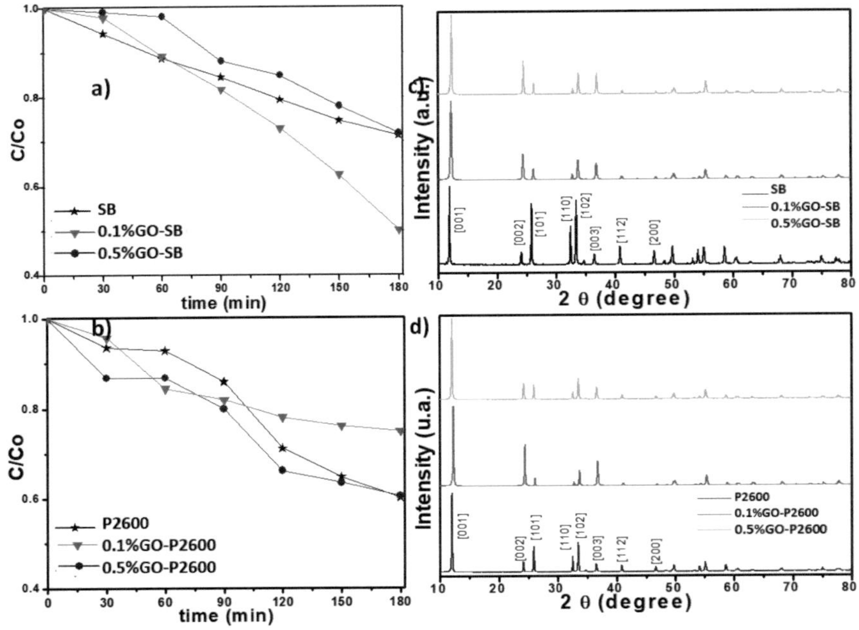

Figure 11. RhB degradation under visible irradiation with BiOCl pure and modified with graphene oxide; OG-SB (a) and OG-P2600 (b). XRD spectra of BiOCl samples; OG-SB (c) and OG-P2600 (d).

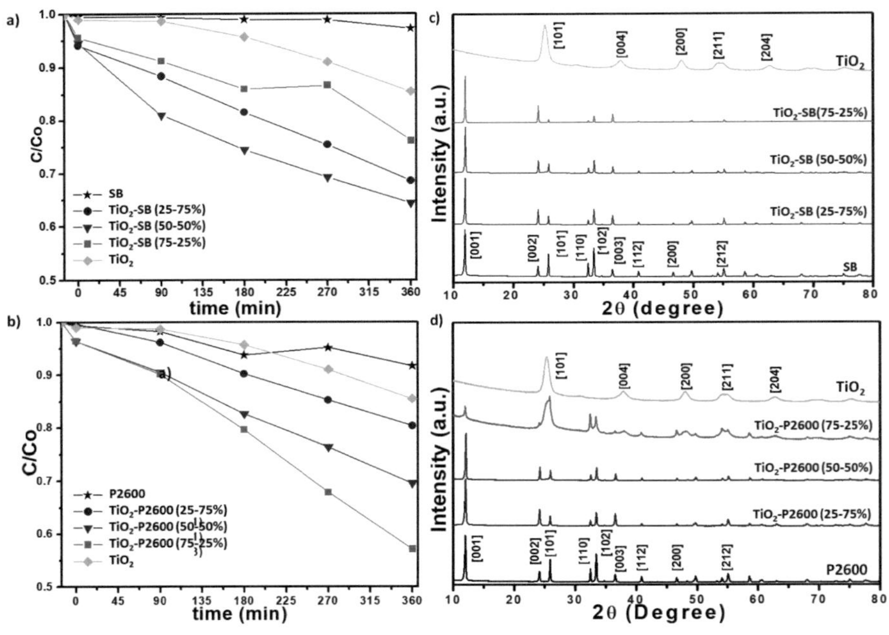

Figure 12. Phenol degradation under visible irradiation with BiOCl pure and modified with titanium dioxide; TiO_2-SB (a) and TiO_2-P2600 (b). XRD spectra of BiOCl samples; TiO_2-SB (c) and TiO_2-P2600 (d).

Author details

Vladimir A. Escobar Barrios[1]*, Dalia Verónica Sánchez Rodríguez[1],
Nancy Ayerim Cervantes Rincón[2] and Alma Berenice Jasso-Salcedo[3]

*Address all correspondence to: vladimir.escobar@ipicyt.edu.mx

1 Instituto Potosino de Investigación Científica y Tecnólogica, A.C. (IPICYT), Advanced Materials Division, San Luis Potosí, México

2 Instituto Potosino de Investigación Científica y Tecnólogica, A.C. (IPICYT), Environmental Sciences Division, San Luis Potosí, México

3 Department of Materials and Environmental Chemistry, Stockholm University, Stockholm, Sweden

References

[1] Lacombe S, Keller N. Photocatalysis: Fundamentals and applications in JEP 2011. Environmental Science and Pollution Research. 2012;**19**:3651-3654. DOI: 10.1007/s11356-012-1040-8

[2] Wang ZL, Kong XY, Ding Y, Gao P, Hughes WL, Yang R, et al. Semiconducting and piezoelectric oxide nanostructures induced by polar surfaces. Advanced Functional Materials. 2004;**14**:943-956. DOI: 10.1002/adfm.200400180

[3] Jasso-Salcedo AB, Hoppe S, Pla F, Escobar-Barrios VA, Camargo M, Meimaroglou D. Modeling and optimization of a photocatalytic process: Degradation of endocrine disruptor compounds by Ag/ZnO. Chemical Engineering Research and Design. 2017;**128**:174-191

[4] Man MT, Kim JH, Jeong MS, Do ATT, Lee HS. Oriented ZnO nanostructures and their application in photocatalysis. Journal of Luminescence. 2017;**185**:17-22. DOI: 10.1016/j.jlumin.2016.12.046

[5] Cantwell G, Harsch WC, Jogai B. Valence-band ordering in ZnO. Physical Review B: Condensed Matter and Materials Physics. 1999;**60**:2340-2344. DOI: 10.1103/PhysRevB.60.2340

[6] Porter F. Modified Metallic Oxides for Efficient Photocatalysis. First edit; 1991

[7] Jia W, Shang Y, Gong L, Chen X. Synthesis of Al-ZnO nanocomposite and its potential application in photocatalysis and electrochemistry. Inorganic Chemistry Communications. 2018;**88**:51-55. DOI: 10.1016/j.inoche.2017.12.013

[8] Schmidt-Mende L, MacManus-Driscoll JL. ZnO–Nanostructures, defects, and devices. Materials Today. 2007;**10**:40-48. DOI: 10.1016/S1369-7021(07)70078-0

[9] Daneshvar N, Aber S, Seyed Dorraji MS, Khataee AR, Rasoulifard MH. Photocatalytic degradation of the insecticide diazinon in the presence of prepared nanocrystalline ZnO

powders under irradiation of UV-C light. Separation and Purification Technology. 2007; **58**:91-98. DOI: 10.1016/j.seppur.2007.07.016

[10] Wang C, Astruc D. Recent developments of metallic nanoparticle-graphene nanocatalysts. Progress in Materials Science. 2018;**94**:306-383. DOI: 10.1016/j.pmatsci.2018.01.003

[11] Bai X, Wang L, Zhu Y. Visible photocatalytic activity enhancement of $ZnWO_4$ by graphene hybridization. ACS Catalysis. 2012;**2**:2769-2778. DOI: 10.1021/cs3005852

[12] Shen Y, Fang Q, Chen B. Environmental applications of three-dimensional graphene-based macrostructures: Adsorption, transformation, and detection. Environmental Science & Technology. 2015;**49**:67-84. DOI: 10.1021/es504421y

[13] McClellan K, Halden RU. Pharmaceuticals and personal care products in archived U.S. biosolids from the 2001 EPA national sewage sludge survey. Water Research. 2010;**44**: 658-668. DOI: 10.1016/j.watres.2009.12.032

[14] Walters E, McClellan K, Halden RU. Occurrence and loss over three years of 72 pharmaceuticals and personal care products from biosolids-soil mixtures in outdoor mesocosms. Water Research. 2010;**44**:6011-6020. DOI: 10.1016/j.watres.2010.07.051

[15] Jasso-Salcedo AB, Palestino G, Escobar-Barrios VA. Effect of Ag, pH, and time on the preparation of Ag-functionalized zinc oxide nanoagglomerates as photocatalysts. Journal of Catalysis. 2014;**318**:170-178

[16] Kislov N, Lahiri J, Verma H, Goswami DY, Stefanakos E, Batzill M. Photocatalytic degradation of methyl orange over single crystalline ZnO: Orientation dependence of photoactivity and photostability of ZnO. 2009;**25**:3310-3315. DOI: 10.1021/la803845f

[17] Santiago-Morales J, Gómez MJ, Herrera-López S, Fernández-Alba AR, García-Calvo E, Rosal R. Energy efficiency for the removal of non-polar pollutants during ultraviolet irradiation, visible light photocatalysis and ozonation of a wastewater effluent. Water Research. 2013;**47**:5546-5556. DOI: 10.1016/j.watres.2013.06.030

[18] Xu F, Chen J, Kalytchuk S, Chu L, Shao Y, Kong D, et al. Supported gold clusters as effective and reusable photocatalysts for the abatement of endocrine-disrupting chemicals under visible light. Journal of Catalysis. 2017;**354**:1-12. DOI: 10.1016/j.jcat.2017.07.027

[19] Dai Y, Yin L. Synthesis and photocatalytic activity of Ag-Ti-Si ternary modified-Bi_2O_3 nanoporous spheres. Materials Letters. 2015;**142**:225-228. DOI: 10.1016/j.matlet.2014.12.013

[20] Niu J, Dai Y, Yin L, Shang J, Crittenden JC. Photocatalytic reduction of triclosan on Au–Cu_2O nanowire arrays as plasmonic photocatalysts under visible light irradiation. Physical Chemistry Chemical Physics. 2015;**17**:17421-17428. DOI: 10.1039/C5CP02244D

[21] Zhao L, Deng J, Sun P, Liu J, Ji Y, Nakada N, et al. Nanomaterials for treating emerging contaminants in water by adsorption and photocatalysis: Systematic review and bibliometric analysis. Science of the Total Environment. 2018;**627**:1253-1263. DOI: 10.1016/j.scitotenv. 2018.02.006

[22] Chen F, An W, Liu L, Liang Y, Cui W. Highly efficient removal of bisphenol A by a three-dimensional graphene hydrogel-AgBr@rGO exhibiting adsorption/photocatalysis synergy. Applied Catalysis B: Environmental. 2017;**217**:65-80. DOI: 10.1016/j.apcatb.2017.05.078

[23] Fu H, Xu T, Zhu S, Zhu Y. Photocorrosion inhibition and enhancement of photocatalytic activity for ZnO via hybridization with C_{60}. Environmental Science & Technology. 2008; **42**:8064-8069. DOI: 10.1021/es801484x

[24] Xu T, Zhang L, Cheng H, Zhu Y. Significantly enhanced photocatalytic performance of ZnO via graphene hybridization and the mechanism study. Applied Catalysis B: Environmental. 2011;**101**:382-387. DOI: 10.1016/j.apcatb.2010.10.007

[25] Xie W, Li Y, Sun W, Huang J, Xie H, Zhao X. Surface modification of ZnO with Ag improves its photocatalytic efficiency and photostability. Journal of Photochemistry and Photobiology A: Chemistry. 2010;**216**:149-155. DOI: http://dx.doi.org/10.1016/j.jphotochem.2010.06.032

[26] Kawano K, Komatsu M, Yajima Y, Haneda H, Maki H, Yamamoto T. Photoreduction of Ag ion on ZnO single crystal. Applied Surface Science. 2002;**189**:265-270

[27] Salaün A, Hamilton JA, Iacopino D, Newcomb SB, Nolan MG, Padmanabhan SC, et al. The incorporation of preformed metal nanoparticles in zinc oxide thin films using aerosol assisted chemical vapour deposition. Thin Solid Films. 2010;**518**:6921-6926

[28] Li B, Cao H. ZnO@graphene composite with enhanced performance for the removal of dye from water. Journal of Materials Chemistry. 2011;**21**:3346-3349. DOI: 10.1039/c0jm03253k

[29] Xiong Z, Zhang LL, Ma J, Zhao XS. Photocatalytic degradation of dyes over graphene-gold nanocomposites under visible light irradiation. Chemical Communications. 2010;**46**: 6099-6101. DOI: 10.1039/c0cc01259a

[30] Jasso-Salcedo AB, Meimaroglou D, Hoppe S, Pla F, Escobar-Barrios VA. Surface modification and immobilization in poly (acrylic acid) of Ag/ZnO for photocatalytic degradation of endocrine-disrupting compounds. Journal of Applied Polymer Science. 2016;**133**:1-12

[31] Leng C, Wei J, Liu Z, Xiong R, Pan C, Shi J. Facile synthesis of PANI-modified $CoFe_2O_4$-TiO_2 hierarchical flower-like nanoarchitectures with high photocatalytic activity. Journal of Nanoparticle Research. 2013;**15**:1643

[32] Zhang H, Zong R, Zhu Y. Photocorrosion inhibition and photoactivity enhancement for zinc oxide via hybridization with monolayer polyaniline. Journal of Physical Chemistry C. 2009;**113**:4605-4611

[33] Lu B, Liu M, Shi H, Huang X, Zhao G. A novel photoelectrochemical sensor for bisphenol A with high sensitivity and selectivity based on surface molecularly imprinted polypyrrole modified TiO_2 nanotubes. Electroanalysis. 2013;**25**:771-779

[34] Yang W, Wen Y, Chen R, Zeng D, Shan B. Study of structural, electronic and optical properties of tungsten doped bismuth oxychloride by DFT calculations. Physical Chemistry Chemical Physics. 2014;**16**:21349-21355. DOI: 10.1039/C4CP02801E

[35] Sun D, Li J, Feng Z, He L, Zhao B, Wang T, et al. Solvothermal synthesis of BiOCl flower-like hierarchical structures with high photocatalytic activity. Catalysis Communications. 2014;**51**:1-4. DOI: 10.1016/j.catcom.2014.03.004

[36] Sarwan B, Pare B, Acharya AD. The effect of oxygen vacancies on the photocatalytic activity of BiOCl nanocrystals prepared by hydrolysis and UV light irradiation. Materials Science in Semiconductor Processing. 2014;**25**:89-97. DOI: 10.1016/j.mssp.2013.09.015

[37] Hao HY, Xu YY, Liu P, Zhang GY. BiOCl nanostructures with different morphologies: Tunable synthesis and visible-light-driven photocatalytic properties. Chinese Chemical Letters. 2015;**26**:133-136. DOI: 10.1016/j.cclet.2014.11.022

[38] Wang Q, Hui J, Huang Y, Ding Y, Cai Y, Yin S, et al. The preparation of BiOCl photocatalyst and its performance of photodegradation on dyes. Materials Science in Semiconductor Processing. 2014;**17**:87-93. DOI: 10.1016/j.mssp.2013.08.018

[39] Gao Y, Wang L, Li Z, Li C, Cao X, Zhou A, et al. Microwave-assisted synthesis of flower-like Ag-BiOCl nanocomposite with enhanced visible-light photocatalytic activity. Materials Letters. 2014;**136**:295-297. DOI: 10.1016/j.matlet.2014.08.026

[40] Zhang Z, Zhou Y, Yu S, Chen M, Wang F. Ag-BiOCl nanocomposites prepared by the oxygen vacancy induced photodeposition method with improved visible light photocatalytic activity. Materials Letters. 2015;**150**:97-100. DOI: 10.1016/j.matlet.2015.03.011

[41] Cheng J, Wang C, Cui Y, Sun Y, Zuo Y, Wang T. Large improvement of visible-light-driven photocatalytic property in AgCl nanoparticles modified black BiOCl microsphere. Materials Letters. 2014;**127**:28-31. DOI: 10.1016/j.matlet.2014.04.012

[42] Xia J, Xu L, Zhang J, Yin S, Li H, Xu H, et al. Improved visible light photocatalytic properties of Fe/BiOCl microspheres synthesized via self-doped reactable ionic liquids. CrystEngComm. 2013;**15**:10132. DOI: 10.1039/c3ce41555d

[43] Hu J, Xu G, Wang J, Lv J, Zhang X, Zheng Z, et al. Photocatalytic properties of Bi/BiOCl heterojunctions synthesized using an in situ reduction method. New Journal of Chemistry. 2014;**38**:4913-4921. DOI: 10.1039/C4NJ00794H

[44] Gao F, Zeng D, Huang Q, Tian S, Xie C. Chemically bonded graphene/BiOCl nanocomposites as high-performance photocatalysts. Physical Chemistry Chemical Physics. 2012;**14**:10572. DOI: 10.1039/c2cp41045a

[45] Kang S, Pawar RC, Pyo Y, Khare V, Lee CS. Size-controlled BiOCl–RGO composites having enhanced photodegradative properties. Journal of Experimental Nanoscience. 2016;**11**:259-275. DOI: 10.1080/17458080.2015.1047420

[46] Tan C, Zhu G, Hojamberdiev M, Okada K, Liang J, Luo X, et al. Co_3O_4 nanoparticles-loaded BiOCl nanoplates with the dominant {001} facets: Efficient photodegradation of organic dyes under visible light. Applied Catalysis B: Environmental. 2014;**152–153**: 425-436. DOI: 10.1016/j.apcatb.2014.01.044

[47] Sun L, Xiang L, Zhao X, Jia CJ, Yang J, Jin Z, et al. Enhanced visible-light photocatalytic activity of BiOI/BiOCl heterojunctions: Key role of crystal facet combination. ACS Catalysis. 2015;**5**:3540-3551. DOI: 10.1021/cs501631n

[48] Liu Z, Xu X, Fang J, Zhu X, Li B. Synergistic degradation of eosin y by photocatalysis and electrocatalysis in UV irradiated solution containing hybrid BiOCl/TiO_2particles. Water, Air, and Soil Pollution. 2012;**223**:2783-2798. DOI: 10.1007/s11270-011-1066-4

[49] Li L, Zhang M, Liu Y, Zhang X. Hierarchical assembly of BiOCl nanosheets onto bicrystalline TiO_2 nanofiber: Enhanced photocatalytic activity based on photoinduced interfacial charge transfer. Journal of Colloid and Interface Science. 2014;**435**:26-33. DOI: 10.1016/j.jcis.2014.08.022

[50] Sánchez-Rodríguez D, Méndez Medrano MG, Remita H, Escobar-Barrios V. Photocatalytic properties of BiOCl-TiO_2 composites for phenol photodegradation. Journal of Environmental Chemical Engineering. 2018;**6**:1601-1612. DOI: 10.1016/j.jece.2018.01.061

[51] Dong S, Pi Y, Li Q, Hu L, Li Y, Han X, et al. Solar photocatalytic degradation of sulfanilamide by BiOCl/reduced graphene oxide nanocomposites: Mechanism and degradation pathways. Journal of Alloys and Compounds. 2016;**663**:1-9. DOI: 10.1016/j.jallcom.2015.12.027

[52] Zhang L, Wang W, Jiang D, Gao E, Sun S. Photoreduction of CO_2 on BiOCl nanoplates with the assistance of photoinduced oxygen vacancies. Nano Research. 2015;**8**:821-831. DOI: 10.1007/s12274-014-0564-2

Chapter 5

Lanthanides Effects on TiO_2 Photocatalysts

Gustavo Lopes Colpani, Micheli Zanetti,
Rubieli Carla Frezza Zeferino, Luciano Luiz Silva,
Josiane Maria Muneron de Mello,
Humberto Gracher Riella, Natan Padoin,
Márcio Antônio Fiori and Cíntia Soares

Additional information is available at the end of the chapter

http://dx.doi.org/10.5772/intechopen.80906

Abstract

Semiconductors have been evaluated to heterogeneous photocatalysis degradation of recalcitrant contaminants in aqueous media due to the capacity of mineralizing these compounds under UV or visible light irradiation. However, this process has the inherent feature of photogenerated charges recombination and the high bandgap energy of the electronic structure of some semiconductors that can reduce the formation of reactive oxygen species, which are responsible for the compound degradation. In this context, structural modifications in semiconductors have been proposed to enhance the photocatalytic activity, such as doping processes with elements that are capable of generating superficial defects that capture the formed electrons, avoiding the recombination, or increasing the density of –OH groups or water molecules on the surface of the catalyst, which can enhance the formation of hydroxyl radicals. Therefore, this brief review is proposed to show the role of lanthanides in the TiO_2 doping and the synthesis method applied, as well as the results discussed in the literature.

Keywords: photocatalyst, lanthanides, TiO_2, doping, optical properties

1. Introduction

Photocatalysis applied in gas or aqueous media involves the reactive oxygen species (ROS) formation, mainly hydroxyl radical (•OH), superoxide ($\bullet O_2^-$), and singlet oxygen ($^1O_2$), which are generated by the oxidation of water molecules or capture of electrons by oxygen.

These species are very effective oxidant agents, and they are capable of degrading recalcitrant compounds due to the high potential of oxidation. These ROS are produced on the semiconductors' surface, such as titanium dioxide (TiO_2), which is considered the most promising material because it is inexpensive, nontoxic, chemically and thermally stable, abundant, and environmentally friendly. However, this semiconductor only will produce these species when it receives a minimum energy amount, called the bandgap energy (E_g), which is capable to remove an electron from the valence band (VB) and transfer it to the conduction band (CB), thus creating the electron-hole pair. For TiO_2, this minimum energy is supplied by photons with frequency in the ultraviolet light region, which possess wavelengths below 400 nm [1–6].

Despite the ability of photocatalytic processes to degrade several compounds by the use of hydroxyl radicals, their use is still not widespread, with scarce industrial applications. This is due to some inherent features of the catalysts employed, such as the recombination of photogenerated charges, which reduces the formation of radicals and, consequently, the efficiency of photocatalytic degradation. Therefore, plenty of studies have been performed to overcome this drawback, like surface modifications that allow the capture of the generated electrons and avoid the recombination. Furthermore, the bandgap energy required for the formation of the electron/vacancy pair, which for TiO_2 is equal to 3.2 eV, restricts this catalyst to the use of a light source that has wavelengths in the ultraviolet region (≈390 nm). Therefore, only 5% of sunlight can be used in photocatalysis that applies this semiconductor, which makes the process more expensive. In this way, superficial modification through metal and nonmetal doping is fundamental to overcome these drawbacks [7–10].

Among the possible dopants, rare earths have been investigated for their ability to increase photocatalytic activity, possibly by reducing bandgap energy due to the introduction of orbitals between the conduction and valence bands, generating impurity energy levels in the semiconductor elements. These states are generated from the 4f level, which are electron deficient. Another hypothesis for this increase in contaminant degradability is that the adsorption of these lanthanides on the surface of the semiconductors generates an imbalance of surface charges, which can produce surface defects and vacancies of oxygen and titanium. These two propositions lead to states that serve as electron scavengers and reduce the recombination of photogenerated charges, increasing the probability of •OH formation. Moreover, the lanthanides adsorbed on the surface may act as traps to water molecule and –OH (hydroxyl anion) groups, which increase their density on the photocatalyst surface and can promote more intense formation of hydroxyl radicals. Another important feature related to the use of these compounds is that they serve as a Lewis base, which could concentrate the contaminants dispersed in aqueous medium on the semiconductor surface enhancing the electron transfer for the direct degradation of the contaminant or increasing the probability of interaction between the molecules and the radicals formed [11–15].

The mechanisms of lanthanides doping on TiO_2 proposed by different researchers have not been the same, with two ways most discussed. Rare earths can be included in the TiO_2 lattice by direct linking or substitution producing a≡Ti–O–Ln–O–Ti≡arrangement, which cause distortions/defects in the lattice due to the mismatch of ionic radius of Ln^{3+} and Ti^{4+}. On the other hand, some doping methods do not supply enough energy to promote the ions substitution in the lattice, and the lanthanides are located in the TiO_2 surface or boundaries of the semiconductor particles, creating Ti–O–Ln bonds [16–20].

Different methods are currently used for doping, like solvothermal, microemulsion, electrospinning, cathodic sputtering, hydrothermal, microwave, sonochemical, and sol–gel, which is the most usual process. However, these methods usually employ reactants such as tetrabutoxide and titanium isopropoxide, which are aggressive to the environment, besides using severe conditions of temperature and pressure. Therefore, the methodology using ultrasound appears as an alternative to avoid the application of secondary reactants and extreme conditions [21–25].

Lanthanides have particular features when compared with other metal oxides dopants, because they inhibit the transition from the anatase phase to the rutile phase, reduce the crystalline size, can easily react with the functional groups of Lewis bases to mineralize organic compounds, and bring luminescent properties to the titanium dioxide [19, 26, 27].

Thus, titanium dioxide doped with lanthanides, which avoid the recombination mechanisms, has been studied in the sense of optimizing the capacity of radical formation due to the increased numbers of defects and oxygen vacancies on the surface of the lattice to trap $e-/h+$ pairs and reduce the recombination, besides that increase the density of –OH groups or water molecules attached to the surface of the catalyst. However, the excess of these dopants reduces the photocatalytic capacity of the semiconductor, possibly due to the excess of vacancies generated, which act as recombination centers instead of electrons scavenger [28–30].

2. TiO_2

The crystalline structures of titanium dioxide are anatase, rutile, and brookite, which is difficult to be synthesized in laboratory. Therefore, the first two lattices are the most prominent in researches. All crystal lattices are composed of TiO_6 octahedra, but with distinct connections between these structures, since anatase has four connections between the edges, brookite has three connections, and rutile has only two. This structure confers to rutile a greater thermodynamic stability among the other polymorphs, as established by the third Pauling rule [31–33].

There is a great interest in the application of titanium dioxide as a photocatalyst because this compound is nontoxic, economical, chemically inert, and photostable to corrosion, besides high thermal stability, intense photocatalytic activity, and strong oxidant power. However, not all crystalline structures have the same efficiency in the absorption of light for catalysis; for example, although rutile is the polymorph thermodynamically more stable, its photocatalytic activity is lower than anatase, possibly due to the high temperature required for its preparation, which causes an increase in particle size, a high rate of electron/vacancy recombination that reduces the number of hydroxyl groups on the surface, and lower electron mobility in relation to anatase [34–37].

However, the rutile phase has a bandgap equal to 3.0 eV, having the ability to absorb radiation in the visible light spectrum, while the anatase has a bandgap of approximately 3.2 eV. Therefore, the interaction of these two crystalline structures often has better photocatalytic results than when they are purely applied, which is possibly due to the visible light absorption by the rutile, which serves as a photosensitizer for the anatase phase. An example of this phase interaction is the commercial P25® titanium dioxide from Degussa Evonik, which has a composition of 80% anatase and 20% rutile, exhibiting high photocatalytic activity [32, 38, 39].

Regarding to the electronic structure, this catalyst can be considered a semiconductor with indirect bandgap for anatase phase, meaning that the valence and conduction bands of this lattice do not have the same maximum and minimum momentum, respectively. This condition leads to a recombination of the electron/vacancy pair that only occurs if assisted by a phonon, resulting in a longer lifetime of this electron generated. Furthermore, TiO_2 is an n-type semiconductor; in other words it has a greater density of electrons than vacancies produced by oxygen, because they are compensated by the presence of Ti^{3+}, which brings the Fermi level closer to the conduction band [40–42].

Bandgap energy for the structure with the highest photocatalytic activity is approximately 3.2 eV, and the valence band, formed mainly by the 2p orbital of the oxygen, has an energy level of approximately 2.60 V, which is more positive than the oxidation potential of water to hydroxyl radicals (E_0 = 2.27 V). In the case of the conduction band, which is constituted mainly by titanium 3d orbitals, the energy level is equal to −0.51 V, with a value that is negatively greater than the oxygen reduction potential of the superoxide radical (E_0 = −0.33 V). As depicted in **Figure 1**, with these conditions titanium dioxide can degrade organic compounds through activation by UV light, which makes it a prominent catalyst [13, 32, 43–45].

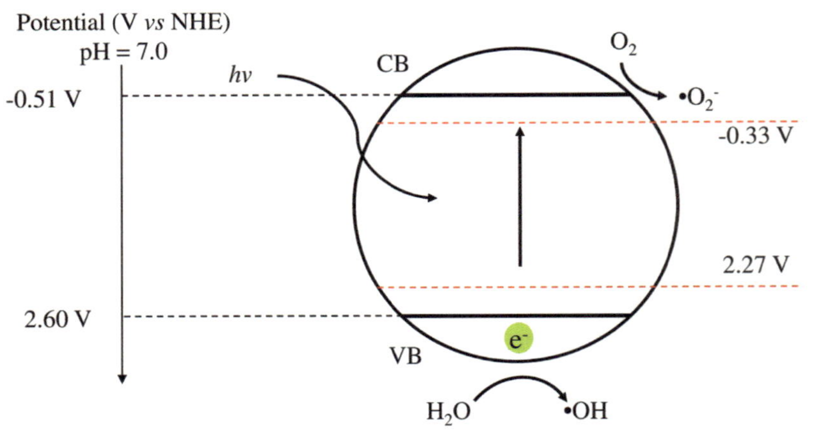

Figure 1. Redox potential of conduction and valence bands and the radical formation.

3. Lanthanides

The rare earths (RE) are a group of elements with physicochemical properties that are very similar, which is constituted by lanthanides (lanthanum to lutetium), scandium, and yttrium, totalizing 17 elements of the periodic table. This series can be separated into light rare earths that comprise low atomic mass elements (lanthanum to europium) and heavy rare earths, which are constituted from lutetium to gadolinium, besides yttrium [19, 46].

These elements have shown great potential for industrial applications because of their unique magnetic, optical, and/or redox properties, with important uses in the field of catalysis, high temperature superconductors, hybrid cars, permanent magnets, nuclear magnetic resonance, rechargeable batteries, manufacture of glass and ceramic materials, shift reagents, etc. The electronic configuration of these chemical elements assure these properties due to the distribution of the electrons with the complete levels until 5, such as xenon, but with the level *f*, it was incomplete and protected by orbitals 6 *s* and/or 5*d*, except for of the scandium and yttrium. Thus, the representation of the electronic configuration is described as [Xe] $4f^n6s^2$ or [Xe] $4f^n5d^16s^2$ for lanthanides (Ln), which usually have an oxidation state equal to 3+ (Ln^{3+}) [47–51].

The orbitals 4*f* that were partially completed are responsible for the optical properties of rare earths, since different arrangements of these orbitals generate different levels of energy, allowing the absorption in a broad spectrum of light radiation, which can vary from ultraviolet to visible. Moreover, the luminescence of some rare earth ions arises from the *f-f* electronic transitions within their partially filled 4*f* orbitals that can occur by electric or magnetic dipole [52, 53].

Another advantage of this configuration is that these orbitals are sterically shielded from the surrounding microenvironment by the filled 5s and 5p orbitals, meaning that there are almost no perturbations of these transitions by other bonding elements, which assure that the optical properties do not undergo sudden changes [48, 54, 55].

4. Surface modification by lanthanides

The surface modification of semiconductors that occurs by the addition of tiny and controlled impurities is denominated doping, and this process is applied to achieve distinct electronic properties from those raw materials. Metal doping has a main feature, the bandgap reduction, which optimizes the ability of a semiconductor to absorb wavelengths with less energy, reduce recombination of the vacancy/electron pairs, and modify the adsorptive capacity of the surface of this photocatalyst. Bandgap reduction occurs possibly due to the formation of energy levels between the conduction and valence bands, such as when transition metals are used, because these incorporate levels 3d in the TiO_2 lattice, which leads to the formation of electronic levels occupied near to the conduction band. However, visible light absorption by semiconductor doped with noble metals, such as gold and silver, may be associated with the surface resonance effect of Plasmon, which is described as an oscillation of the electrons on the surface of these metals, which cause a separation of charge carriers [13, 44, 56].

Another advantage of metal doping is the decrease in photogenerated charges recombination, which is due to the Schottky barrier formation in the interface of semiconductor and metal. The electrons formed migrate, and they are captured by the metal particles, which become active sites for oxygen reduction. This barrier is formed by the electronic binding between the lanthanides and titanium dioxide, allowing the mobility of electrons formed in the semiconductor valence band to the metal, until equilibrium occurs between the Fermi levels [57, 58].

The rare earths when applied in TiO_2 doping have shown one or both features above discussed, mainly due to the presence of incomplete 4f levels, surface defects, and oxygen vacancies. The first ones reduce the TiO_2 bandgap due to the appearance of orbitals between the conduction and valence bands, which generates impure energy levels. This reduction provides an electron transfer from VB until the orbital created and/or from this to the CB requires less energy. In addition, these orbitals *f* have the ability to produce complexes with several Lewis bases, increasing the concentration of these species on the surface of the catalyst and, consequently, the photocatalytic activity. Moreover, these orbitals have the ability to trap the photogenerated electrons due to their incomplete levels, which avoids the recombination of the charge carriers [11–15].

Another advantage of lanthanides is the ability of some ions of these elements (Ln^{3+}) to generate surface defects and titanium and oxygen vacancies. This occurs possibly because some lanthanides ions have an atomic radius higher than Ti^{4+} (0.68 Å), such as La^{3+} (1.15 Å) and Sm^{3+} (1.24 Å). In these cases, Ln^{3+} is not able to replace the titanium ions in the semiconductor lattice, and these ions only are adsorbed on the surface, forming Ti–O–Ln^{3+} bonds. This fact leads to an imbalance in the surface charges of the TiO_2 crystal structure. Since titanium atom has a Pauling electronegativity value (1.54) higher than those presented by lanthanum (1,10) and samarium atoms (1,17), a transfer of electrons occur from these lanthanides to the titanium, which is converted from Ti^{4+} to Ti^{3+}. The excess of negative charge disturbs the electronegativity, and it is necessary remove an ion of O^{2-} for every two Ti^{3+} ions produced, which causes the formation of oxygen and titanium vacancies, generating a surface defect. Furthermore, the presence of lanthanides, dispersed as interstitial impurities on the TiO_2 lattice causes surface defects. The presence of these vacancies and defects causes the capture of electrons produced on the surface of the photocatalyst and decreases the recombination of photogenerated charge carriers. However, an excess of lanthanides on the photocatalyst surface can lead to a high density of defects and vacancies, which creates recombination centers [30, 59–64].

Rare earth application as dopant also can increase the surface area of the catalyst because it reduces the growth of the crystal structure. Besides, these elements can increase the thermal stability, since they interfere in the conversion of the anatase phase to rutile during the calcination of the doped semiconductors [12, 25, 65].

Lanthanides have been applied as TiO_2 dopants in many researches, highlighting yttrium, cerium, neodymium, praseodymium, europium, samarium, and lanthanum, with promising

http://dx.doi.org/10.5772/intechopen.80906

photocatalytic activity responses, such as studies performed by Liu et al. [66] which evaluated the cerium doping on titanium dioxide using chemical coprecipitation method, with molar ratio from 0.5 to 2.5% of cerium in relation to the semiconductor to evaluate the degradation of methylene blue 2,3-dichlorophenol and benzene, under visible and UV light. The best results occur when the RE concentration was equal to 0.50% (mol•mol^{-1}). Xue et al. [67] evaluated the effects of cerium doping into TiO_2 nanotubes by chemical impregnation and ultrasound. The molar ratio applied was 0.1–1.0% of the lanthanide in relation to the semiconductor, with subsequent calcination of all samples at 673 K. The results obtained in the degradation of glyphosate under UV light showed that the molar ratio equal 0.15% optimizes the photocatalysis.

Cerium doping was also recently evaluated by Pei et al. [68] in the titanium dioxide nanorods formation. The inclusion of cerium in the semiconductor occurred by hydrothermal method. The inclusion of this element changed the semiconductor wavelength absorption to 480 nm, with bandgap energy equal to 2.65 eV, meaning that photocatalysis can occur in visible light.

Neodymium (Nd) has attracted interest as a semiconductor dopant because of its ability to absorb and emit wavelengths in the range from UV to IR radiation, besides the usual features of RE, such as discussed above. Du et al. [69] used this element in the titanium dioxide film doping by sol–gel method. In this research, the photocatalytic effects of the mass ratios of neodymium (0.1 and 0.9%) after calcination at 823 K for 2 h were evaluated. The obtained films were used in the degradation of methylene blue under UV light, with efficiency upper of 92% in the removal of this dye when using a mass ratio of 0.10%. Possibly, this increment occurs due to a reduction in bandgap energy, since analysis showed an increase in the visible light wavelengths absorption by the doped semiconductor. Another important research was done by Thomas et al. [70] who applied molar ratios equal to 0.30, 0.50, and 0.70% of neodymium to doping titanium dioxide via sol-gel method. It evaluated the insertion of a third component, phosphotungstic acid, in the degradation of methylene blue and 4-chlorophenol under visible light. The results obtained showed an increase in the degradation in relation to the pure semiconductor, with the best result when the molar ratio was equal to 0.50%.

Another element evaluated for titanium dioxide doping is praseodymium, classified as a light rare earth with molar mass equal to 140.9 g·mol^{-1}. Liang et al. [71] evaluated the degradation of the reactive yellow dye 4 under UV and visible radiation using TiO_2 doped with different ionic ratios of praseodymium (0.50–1.80%), which was included in the crystal lattice by sol–gel method, with calcination at 773 K for 2 h. The photodegradation results demonstrated that Ln^{3+}:TiO_2 equal to 1.0 and 1.5% optimizes the photocatalytic capacity of the semiconductor in degrading the contaminant under UV and visible light, respectively. This increase in the catalytic activity of praseodymium doped semiconductors was also confirmed according to research published by Kralova et al. [72], which demonstrated that titanium dioxide doped with 0.3 mol% of praseodymium at different temperatures (723, 823 and 923 K) and calcination times (4, 8 and 12 h) promotes a reduction in bandgap energy (from 3.20 to 3.14 eV) when the photocatalyst was calcined at 723 K for 8 h. The efficiency increase in the degradation of an organochlorine pesticide under UV light (LED) possibly is related to milder temperatures joined with intermediate times, which ensure a diffusion of the lanthanide without changes in the crystalline phase.

The studies performed by Leal et al. [73] are prominent due to the extensive evaluation about lanthanides doping of TiO_2 lattices. The light lanthanide series (La-Eu), besides gadolinium, were evaluated for the degradation of orange and violet methyl by UV light. The pH values (3.1 and 5.6) and the mass fraction of Ln^{3+}:TiO_2 (0.1 and 0.3%) were used as parameters for the experiments, with semiconductor doping by the sol–gel method and calcination at 773 K for 2 h. The results showed an increase (approximately 37%) in the degradation of the compounds when the doped semiconductors were applied instead of pure TiO_2. Therefore, this research provides an overview of the potential application of these lanthanides to degrade different compounds under different process conditions.

Reszczyńska et al. [74] also evaluated the effects of the application of yttrium, praseodymium, erbium, and europium in the photodegradation efficiency of phenol, under visible and UV light. Furthermore, hydrothermal and sol-gel methods for semiconductor doping were compared in relation to photocatalytic activity response. It used the molar ratios of 0.25 and 0.50% of Ln^{3+} in relation to titanium dioxide. The results showed an increase in phenol degradation under visible light for doped photocatalysts when compared to pure TiO_2. When UV light was applied as radiation source only, the hydrothermal process showed better results than the pure semiconductor. However, if the decomposition efficiencies of the contaminant are compared, it is remarkable that the hydrothermal process overcomes the sol-gel method in most cases. This possibly is related to the high surface area, low crystallite dimensions, and higher density of –OH obtained in the first method employed. In addition, the lower concentration of rare earths in doping showed the best degradation results of the molecule evaluated.

Lanthanum (La) doping has an extensive discussion about the mechanisms, which are well-known. However, different methods, as well as the optimized rare earth amount for each of these methods, are still evaluated, without being exhausted. Among the researches carried out, Li et al. [75] evaluated different amounts (0–0.50%) of lanthanum in the TiO_2 doping, which are impregnated in the semiconductor by sol-gel method, with subsequent calcination (973–1173 K). The best response in methyl orange photodegradation occurred when the concentration of lanthanum was 0.05% with a calcination temperature at 973 K.

Jun et al. [76] analyzed the photocatalytic activity of TiO_2 doped with different concentrations of lanthanum (from 0 to 0.90% $w \bullet w^{-1}$) by sol–gel method followed by calcination at 823 K for 2 h. The results showed that mass ratio 0.30% (0.17% $mol \bullet mol^{-1}$) demonstrated the best results in methylene blue removal. It was proposed that the increase in photocatalytic efficiency is due to the formation of vacancies and defects from the presence of lanthanide on the TiO_2 surface. It was also possible to verify that an increase in lanthanum concentration causes a reduction in photocatalytic efficiency, possibly because La serves as a mediator in the interfacial charge transfer or as a recombination center.

Samarium (Sm) has been used to doping semiconductors because its presence causes significant improvements in the degradation of compounds, which leads the semiconductors to absorb wavelengths in the visible light spectrum, besides low cost of the compound when compared with others lanthanides. Tang et al. [77] doped titanium dioxide with different concentrations of samarium (0–2.16% $mol \bullet mol^{-1}$) by the sol-gel method followed by different calcination temperatures (623–1123 K). The pollutants evaluated were methanol and acetone, with best results when the lowest concentrations of samarium (0.30 and 0.43%) were applied. The photocatalytic degradations achieved results of upper 90% when applied by UV light, which demonstrates that an increase in the amount of rare earths can lead these photocatalysts from electron traps to recombination centers.

Xiao et al. [78] demonstrated that methylene blue degradation under UV light showed best responses when the lowest samarium molar ratio was applied (0.50%). The sol-gel method was used for the inclusion of Sm in the TiO_2 lattice, and the samples were calcined for 2 h at 873 K.

4.1. Calcination temperature effects

Calcination is a vital step for doping, since it allows the activation and/or fixation of the dopant in the semiconductor crystal structure, besides the removal of impurities and the increase in the density of vacancies due to the removal of oxygen from the photocatalyst lattice, in addition to promoting an increase in crystallization. However, an excessive increase in the calcination temperature can lead to a particle aggregation and, consequently, reduction of the surface area, besides the conversion of the anatase phase to rutile, which can affect the photocatalytic activity. Therefore, the calcination temperature control is essential to assure high photocatalytic activity [75, 79–81].

Usually anatase to rutile phase transformation occurs at temperatures between 500 and 750°C, as soon as there is an increase in the crystalline size of anatase when the calcination temperature enhances. However, the lanthanides doping shift the phase transformation to higher temperatures (above 700°C) and suppress the anatase crystalline growth between 500 and 700°C [82–86].

Chen et al. [87] evaluated the effect of rare earth doping (0.20, 0.50, 1.0, and 2.0% mol) by hydrothermal method and the calcination temperature (673, 773, 873 and 1073 K). The results obtained showed that the best photocatalytic activity was achieved when the temperature of calcination was 773 K for 2 h, with a crystallite size equal to 15 nm. It was possible to verify that an increase of the temperature reduces the surface area of the crystal structures formed.

Cruz et al. [88] published a research about titanium dioxide doped with samarium via sol–gel method, in which the photocatalytic degradation of a herbicide under UV light was investigated. The parameters evaluated were the calcination temperature and the samarium concentration, with the best efficiency in herbicide degradation when the samples were calcined at 773 K for 4 h. Similar results were obtained by Yang et al. [76], which showed that the degradation of the methylene blue by TiO_2 doped with neodymium and fluorine via sol-gel method was optimized when the temperature and time of calcination were equal to 773 K and 3 h, respectively.

Li et al. [89] also evaluate the effect of temperature on the photocatalytic activity of titanium dioxide doped with europium via hydrothermal method. It used calcination temperatures between 573 and 1173 K. The temperature at 773 K, with a calcination time of 4 h, improves the degradability of the contaminant, probably because of the increase in crystallization and reduction of defects, which improve the ability to absorb visible light wavelengths.

5. Conclusion

The degradation of recalcitrant compounds using semiconductors doped with lanthanides, in aqueous media, is a promising area due to the increasing incidence of these compounds in water media due to the inefficiency of current processes in eliminating such molecules. Doping semiconductors with lanthanides has been an efficient strategy to generate defects in the surfaces of the materials. These defects act as active sites that capture the electrons and prevent recombination events or increase the density of hydroxyl groups and increase the amount of water molecules bound on the surface of photocatalytic materials. All these mechanisms increase the efficiency of the photodegradation processes. However, the excess of these dopants reduces the photocatalytic capacity of the semiconductor, possibly due to the excess of vacancies generated, which act as recombination centers instead of electrons scavenger.

Acknowledgements

The authors acknowledge the Community University of Chapecó and Region (Unochapecó), for technical and scientific support.

Conflict of interest

There are no conflicts to declare.

Author details

Gustavo Lopes Colpani[1]*, Micheli Zanetti[1], Rubieli Carla Frezza Zeferino[1], Luciano Luiz Silva[1], Josiane Maria Muneron de Mello[1], Humberto Gracher Riella[2], Natan Padoin[2], Márcio Antônio Fiori[1] and Cíntia Soares[2]

*Address all correspondence to: g_colpani@unochapeco.edu.br

1 Community University of Chapecó Region – Unochapecó, Chapecó, Brazil

2 Federal University of Santa Catarina – UFSC, Florianópolis, Brazil

References

[1] Devi LG, Kumar SG. Exploring the critical dependence of adsorption of various dyes on the degradation rate using Ln^{3+}-TiO_2 surface under UV/solar light. Applied Surface Science. 2012;**261**:137-146. DOI: 10.1016/j.apsusc.2012.07.121

[2] Leong S, Ramzjou A, Wang K, Hapgood K, Zhang X, Wang H. TiO_2 based photocatalytic membranes: A review. Journal of Membrane Science. 2014;**472**:167-184. DOI: 10.1016/j.memsci.2014.08.016

[3] Fotiou T, Triantis TM, Kaloudis T, O'shea KE, Dionysiou DD, Hiskia A. Assessment of the roles of reactive oxygen species in the UV and visible light photocatalytic degradation of cyanotoxins and water taste and odor compounds using C–TiO_2. Water Research. 2016;**90**:52-61. DOI: 10.1016/j.watres.2015.12.006

[4] Ribao P, Corredor J, Rivero MJ, Ortiz I. Role of reactive oxygen species on the activity of noble metal-doped TiO_2 photocatalysts. Journal of Hazardous Materials. DOI: 10.1016/j.jhazmat.2018.05.026. In press

[5] Khan SA, Khan SB, Asiri AM. Layered double hydroxide of Cd-Al/C for the mineralization and de-coloration of dyes in solar and visible light exposure. Scientific Reports. 2016;**6**(35107):1-14. DOI: 10.1038/srep35107

[6] Arshad T, Khan SA, Faisal M, Shah Z, Akhtar K, Asiri AM, et al. Cerium based photocatalysts for the degradation of acridine orange in visible light. Journal of Molecular Liquids. 2017;**241**:20-26. DOI: 10.1016/j.molliq.2017.05.079

[7] Singh S, Mahalingam H, Singh PK. Polymer-supported titanium dioxide photocatalysts for environmental remediation: A review. Applied Catalysis A: General. 2013;**462-463**: 178-195. DOI: 10.1016/j.apcata.2013.04.039

[8] Zhao C, Pelaez M, Dionysiou DD, Pillai SC, Byrne JA, O'shea KE. UV and visible light activated TiO_2 photocatalysis of 6-hydroxymethyluracil, a model compound for the potent cyanotoxin cylindrospermopsin. Catalysis Today. 2014;**224**:70-76. DOI: 10.1016/j.cattod.2013.09.042

[9] Etacheri V, Valentin C, Schneider J, Bahnemann D, Pillai SC. Visible-light activation of TiO_2 photocatalysts: Advances in theory and experiments. Journal of Photochemistry and Photobiology, C: Photochemistry Reviews. 2015;**25**:1-29. DOI: 10.1016/j.jphotochemrev.2015.08.003

[10] Wang W, Huang G, Yu JC, Wong PK. Advances in photocatalytic disinfection of bacteria: Development of photocatalysts and mechanisms. Journal of Environmental Sciences. 2015;**34**:232-247. DOI: 10.1016/j.jes.2015.05.003

[11] Kumaresan L, Prabhu A, Palanichamy M, Arumugam E, Murugesan V. Synthesis and characterization of Zr^{4+}, La^{3+} and Ce^{3+} doped mesoporous TiO_2: Evaluation of their photocatalytic activity. Journal of Hazardous Materials. 2011;**186**:1183-1192. DOI: 10.1016/j.jhazmat.2010.11.124

[12] Cao Y, Zhao Z, Yi J, Ma C, Zhou D, Wang R, et al. Luminescence properties of Sm^{3+}-doped TiO_2 nanoparticles: Synthesis, characterization, and mechanism. Journal of Alloys and Compounds. 2013;**553**:12-20. DOI: 10.1016/j.jallcom.2012.11.149

[13] Coronado JM, Fresno F, Hernández-Alonso MD, Portela R. Design of Advanced Photocatalytic Materials for Energy and Environmental Applications, Green Energy and Technology. 1st ed. London: Springer-Verlag; 2013. DOI: 10.1007/978-1-4471-5061-9_1. 351p

[14] Liu Y, Fang P, Cheng Y, Gao Y, Chen F, Liu Z, et al. Study on enhanced photocatalytic performance of cerium doped TiO_2-based nanosheets. Chemical Engineering Journal. 2013;**219**:478-485. DOI: 10.1016/j.cej.2012.12.098

[15] Sudhagar P, Devadoss A, Nakata K, Terashima C, Fujishima A. Enhanced photoelectrocatalytic water splitting at hierarchical Gd^{3+}:TiO_2 nanostructures through amplifying light reception and surface states passivation. Journal of the Electrochemical Society. 2015;**162**(3):H108-H114. DOI: 10.1149/2.0161503jes

[16] Colpani GL, Zanetti JT, Cecchin F, Dal'Toé A, Fiori MA, Moreira RFPM, et al. Carboxymethyl-β-cyclodextrin functionalization of TiO_2 doped with lanthanum: Characterization and enhancement of photocatalytic activity. Catalysis Science & Technology. 2018;**8**:2636-2647. DOI: 10.1039/c7cy02115a

[17] Dal'Toé ATO, Colpani GL, Padoin N, Fiori MA, Soares C. Lanthanum doped titania decorated with silver plasmonic nanoparticles with enhanced photocatalytic activity under UV-visible light. Applied Surface Science. 2018;**441**:1057-1071. DOI: 10.1016/j.apsusc.2018.01.291

[18] Ali KA, Abdullah AZ, Mohamed AR. Visible light responsive TiO_2 nanoparticles modified using Ce and La for photocatalytic reduction of CO_2: Effect of Ce dopant content. Applied Catalysis A: General. 2017;**537**:111-120. DOI: 10.1016/j.apcata.2017.03.022

[19] Mazierski P, Mikolajczyk A, Bajorowicz B, Malankowska A, Zaleska-Medynska A, Nadolna J. The role of lanthanides in TiO_2-based photocatalysis: A review. Applied Catalysis B: Environmental. 2018;**233**:301-317. DOI: 10.1016/j.apcatb.2018.04.019

[20] Bingham S, Daoud WA. Recent advances in making nano-sized TiO_2 visible-light active through rare-earth metal doping. Journal of Materials Chemistry. 2011;**21**:2041-2050. DOI: 10.1039/c0jm02271c

[21] Gomez V, Balu AM, Ruiz JCS, Irusta S, Dionysiou DD, Luque R, et al. Microwave-assisted mild-temperature preparation of neodymium-doped titania for the improved photodegradation of water contaminants. Applied Catalysis A: General. 2012;**441-442**:47-53. DOI: 10.1016/j.apcata.2012.07.003

[22] Shirsath SR, Pinjari DV, Gogate PR, Sonawane SH, Pandit AB. Ultrasound assisted synthesis of doped TiO_2 nano-particles: Characterization and comparison of effectiveness for photocatalytic oxidation of dyestuff effluent. Ultrasonics Sonochemistry. 2013;**20**: 277-286. DOI: 10.1016/j.ultsonch.2012.05.015

[23] Zheng JQ, Zhu YJ, Xu JS, Lu BQ, Qi C, Chen F, et al. Microwave assisted rapid synthesis and photocatalytic activity of mesoporous Nd-doped $SrTiO_3$ nanospheres and nanoplates. Materials Letters. 2013;**100**:62-65. DOI: 10.1016/j.matlet.2013.02.107

[24] Sathishkumar P, Mangalaraja RV, Rozas O, Mansilla HD, Pinilla MAG, Anandan S. Low frequency ultrasound (42 kHz) assisted degradation of Acid Blue 113 in the presence of visible light driven rare earth nanoclusters loaded TiO_2 nanophotocatalysts. Ultrasonics Sonochemistry. 2014;**21**:1675-1681. DOI: 10.1016/j.ultsonch.2014.03.004

[25] Yaoguang Y, Gang C, Yansong Z, Zhonghui H. Recent advances in rare earth elements modification of inorganic semiconductor based photocatalysts for efficient solar energy conversion: A review. Journal of Rare Earths. 2015;**33**(5):453-462. DOI: 10.1016/S1002-0721(14)60440-3

[26] Shayegan Z, Lee C, Haghighat F. TiO_2 photocatalyst for removal of volatile organic compounds in gas phase—A review. Chemical Engineering Journal. 2018;**334**:2408-2439. DOI: 10.1016/j.cej.2017.09.153

[27] Zheng X, Li X, Peng H, Wen J. Ag-decorated core-shell Sm_2O_3@TiO_2 nanocomposites with enhanced visible-light photocatalytic performance. Journal of Physics and Chemistry of Solids. 2018;**123**:206-215. DOI: 10.1016/j.jpcs.2018.07.022

[28] Jaimy KB, Ghosh S, Warrier KG. Enhanced visible light activity of nano-titanium dioxide doped with multiple ions: Effect of crystal defects. Journal of Solid State Chemistry. 2012;**196**:465-470. DOI: 10.1016/j.jssc.2012.06.048

[29] Samadi M, Zirak M, Naseri A, Khorashadizade E, Moshfegh AZ. Recent progress on doped ZnO nanostructures for visible-light photocatalysis. Thin Solid Films. 2016;**605**:2-19. DOI: 10.1016/j.tsf.2015.12.064

[30] Nešić J, Manojlović DD, Anđelković I, Dojčinović BP, Vulić PJ, Krstić J, et al. Preparation, characterization and photocatalytic activity of lanthanum and vanadium co-doped mesoporous TiO_2 for azo-dye degradation. Journal of Molecular Catalysis A: Chemical. 2013;**378**:67-75. DOI: 10.1016/j.molcata.2013.05.018

[31] Marschall R, Wang L. Non-metal doping of transition metal oxides for visible-light photocatalysis. Catalysis Today. 2014;**225**:111-135. DOI: 10.1016/j.cattod.2013.10.088

[32] Verbruggen SW. TiO_2 photocatalysis for the degradation of pollutants in gas phase: From morphological design to plasmonic enhancement. Journal of Photochemistry and Photobiology, C: Photochemistry Reviews. 2015;**24**:64-82. DOI: 10.1016/j.jphotochemrev.2015.07.001

[33] Patil RA, Devan RS, Liou Y, Ma Y. Efficient electrochromic smart windows of one-dimensional pure brookite TiO_2 nanoneedles. Solar Energy Materials & Solar Cells. 2016;**147**:240-245. DOI: 10.1016/j.solmat.2015.12.024

[34] Nakata K, Fujishima A. TiO_2 photocatalysis: Design and applications. Journal of Photochemistry and Photobiology, C: Photochemistry Reviews. 2012;**13**:169-189. DOI: 10.1016/j.jphotochemrev.2012.06.001

[35] Lan Y, Lu Y, Ren Z. Mini review on photocatalysis of titanium dioxide nanoparticles and their solar applications. Nano Energy. 2013;**2**:1031-1045. DOI: 10.1016/j.nanoen.2013.04.002

[36] Gaya UI. Heterogeneous Photocatalysis Using Inorganic Semiconductor Solids. Inglaterra: Springer Science+Business Media Dordrecht; 2014. DOI: 10.1007/978-94-007-7775-0

[37] Zangeneh H, Zinatizadeh AAL, Habibi M, Akia M, Isa MH. Photocatalytic oxidation of organic dyes and pollutants in wastewater using different modified titanium dioxides: A comparative review. Journal of Industrial and Engineering Chemistry. 2015;**26**:1-36. DOI: 10.1016/j.jiec.2014.10.043

[38] Krishna MV, Madhavi G, Idris NF, Idris SAM, Chowdary LRK. Photocatalysis of β-blockers—An overview. Arabian Journal of Chemistry. 2014. DOI: 10.1016/j.arabjc.2014.10.044. In press

[39] Monteiro RAR, Miranda SM, Silva CR, Faria JL, Silva AMT, Boaventura RAR, et al. Gas phase oxidation of n-decane and PCE by photocatalysis using an annular photoreactor packed with a monolithic catalytic bed coated with P25 and PC500. Applied Catalysis B: Environmental. 2015;**165**:306-315. DOI: 10.1016/j.apcatb.2014.10.026

[40] Ângelo J, Andrade L, Madeira LM, Mendes A. An overview of photocatalysis phenomena applied to NO_x abatement. Journal of Environmental Management. 2013;**129**: 522-539. DOI: 10.1016/j.jenvman.2013.08.006

[41] Miessler GL, Fischer PJ, Tarr DA. Química Inorgânica. 5th ed. São Paulo: Pearson Education do Brasil; 2014. 649p

[42] Zhang J, Zhou P, Liu J, Yu J. New understanding of the difference of photocatalytic activity among anatase, rutile and brookite TiO_2. Physical Chemistry Chemical Physics. 2014;**16**:20382-20386. DOI: 10.1039/C4CP02201G

[43] Henderson MA. A surface science perspective on TiO_2 photocatalysis. Surface Science Reports. 2011;**66**:185-297. DOI: 10.1016/j.surfrep.2011.01.001

[44] Park H, Park Y, Kim W, Choi W. Surface modification of TiO_2 photocatalyst for environmental applications. Journal of Photochemistry and Photobiology, C: Photochemistry Reviews. 2013;**15**:1-20. DOI: 10.1016/j.jphotochemrev.2012.10.001

[45] Wen J, Li X, Liu W, Fang Y, Xie J, Xu Y. Photocatalysis fundamentals and surface modification of TiO_2 nanomaterials. Chinese Journal of Catalysis. 2015;**36**:2049-2070. DOI: 10.1016/S1872-2067(15)60999-8

[46] Jha MK, Kumari A, Panda R, Kumar JR, Yoo K, Lee JY. Review on hydrometallurgical recovery of rare earth metals. Hydrometallurgy. 2016;**161**:77-101. DOI: 10.1016/j.hydromet.2016.01.035

[47] Turra C. Impacto dos elementos terras raras no agrossistema citrícola [Thesis]. São Paulo: Universidade de São Paulo; 2010

[48] Atwood DA. The Rare Earth Elements: Fundamentals and Applications. United Kingdom: John Wiley & Sons, Ltd; 2012. 649p

[49] Lima JF. Nanocompostos a base de cério com aplicações na absorção da radiação ultravioleta [thesis]. São Paulo: Universidade de São Paulo; 2013

[50] Cotton S. Lanthanide and Actinide Chemistry. United Kingdom: John Wiley & Sons Ltd; 2006. DOI: 10.1002/0470010088. 263p

[51] Quirino WG. Produção e caracterização de dispositivos orgânicos eletroluminescentes (OLEDs) baseados em complexos β-dicetonatos de terras-raras [Thesis]. Rio de Janeiro: Pontifícia Universidade Católica; 2007

[52] Huang C. Rare Earth Coordination Chemistry Fundamentals and Applications. Singapore: John Wiley & Sons; 2010. DOI: 10.1002/9780470824870. 575p

[53] Li X, Zhang F, Zhao D. Highly efficient lanthanide upconverting nanomaterials: Progresses and challenges. Nano Today. 2013;**8**:643-676. DOI: 10.1016/j.nantod.2013.11.003

[54] Malta OL. Mechanisms of non-radiative energy transfer involving lanthanide ions revisited. Journal of Non-Crystalline Solids. 2008;**354**:4770-4776. DOI: 10.1016/j.jnoncrysol.2008.04.023

[55] Shimizu Y, Ueda K. Lanthanide 4f energy levels in perovskite-type $YAlO_3$. Journal of Luminescence. 2015;**168**:14-19. DOI: 10.1016/j.jlumin.2015.07.018

[56] Fornari AMD. Atividade fotocatalítica e fotoeletroquímica de nanotubos de TiO_2 impregnados com nanopartículas de metais nobres ou pontos quânticos para aplicação na produção de hidrogênio [Thesis]. Rio Grande do Sul: Universidade Federal do Rio Grande do Sul; 2014

[57] Dong H, Zeng G, Tang L, Fan C, Zhang C, He X, et al. An overview on limitations of TiO_2-based particles for photocatalytic degradation of organic pollutants and the corresponding countermeasures. Water Research. 2015;**79**:128-146. DOI: 10.1016/j.watres.2015.04.038

[58] Ferreira Neto VJM. Processo fotocatalítico aplicado à geração de energia—Redução fotocatalítica de dióxido de carbono [Thesis]. Rio de Janeiro: Universidade Federal do Rio de Janeiro; 2015

[59] Cai H, Chen X, Li Q, He B, Tang Q. Enhanced photocatalytic activity from Gd, La codoped TiO_2 nanotube array photocatalysts under visible-light irradiation. Applied Surface Science. 2013;**284**:837-842. DOI: 10.1016/j.apsusc.2013.08.018

[60] Lan X, Wang L, Zhang B, Tian B, Zhang J. Preparation of lanthanum and boron co-doped TiO_2 by modified sol–gel method and study their photocatalytic activity. Catalysis Today. 2014;**224**:163-170. DOI: 10.1016/j.cattod.2013.10.062

[61] Callister WD, Rethwisch DG. Materials Science and Engineering: An Introduction. 7th ed. New York: John Wiley & Sons; 2007. 721p

[62] Meksi M, Turki A, Kochkar H, Bousselmi L, Guillard C, Berhault G. The role of lanthanum in the enhancement of photocatalytic properties of TiO_2 nanomaterials obtained by calcination of hydrogenotitanate nanotubes. Applied Catalysis B: Environmental. 2016;**181**:651-660. DOI: 10.1016/j.apcatb.2015.08.037

[63] Tao Z, Ding H, Chen X, Hou G, Zhang Q, Tang M, et al. The co-doping effect of Sm and In on ceria for electrolyte application in IT-SOFC. Journal of Alloys and Compounds. 2016;**663**:750-754. DOI: 10.1016/j.jallcom.2015.12.164

[64] Zhou X, Zhang X, Feng X, Zhou J, Zhou S. Preparation of a La/N co-doped TiO_2 film electrode with visible light response and its photoelectrocatalytic activity on a Ni substrate. Dyes and Pigments. 2016;**125**:375-383. DOI: 10.1016/j.dyepig.2015.10.044

[65] Maba G. Rare earth doped titania/carbon nanomaterials composite photocatalysts for water treatment [Thesis]. Johannesburg: University of Johannesburg; 2015

[66] Liu TX, Li XZ, Li FB. Enhanced photocatalytic activity of Ce^{3+}-TiO_2 hydrosols in aqueous and gaseous phases. Chemical Engineering Journal. 2010;**157**:475-482. DOI: 10.1016/j.cej.2009.12.010

[67] Xue W, Zhang G, Xu X, Yang X, Liu C. Preparation of titania nanotubes doped with cerium and their photocatalytic activity for glyphosate. Chemical Engineering Journal. 2011;**167**:397-402. DOI: 10.1016/j.cej.2011.01.007

[68] Pei LZ, Liu HD, Lin N, Yu HY. Hydrothermal synthesis of cerium titanate nanorods and its application in visible light photocatalysis. Materials Research Bulletin. 2015;**61**:40-46. DOI: 10.1016/j.materresbull.2014.09.094

[69] Du J, Qi W, Shan Z, Xin G, Jiao L, Haizhi G, et al. Effect of hydroxyl groups on hydrophilic and photocatalytic activities of rare earth doped titanium dioxide thin films. Journal of Rare Earths. 2015;**33**(2):148-153. DOI: 10.1016/S1002-0721(14)60395-1

[70] Thomas J, Radhika S, Yoon M. Nd^{3+}-doped TiO_2 nanoparticles incorporated with heteropoly phosphotungstic acid: A novel solar photocatalyst for degradation of 4-chlorophenol in water. Journal of Molecular Catalysis A: Chemical. 2016;**411**:146-156. DOI: 10.1016/j.molcata.2015.10.021

[71] Liang C, Liu C, Li F, Wu F. The effect of praseodymium on the adsorption and photocatalytic degradation of azo dye in aqueous Pr^{3+}-TiO_2 suspension. Chemical Engineering Journal. 2009;**147**:219-225. DOI: 10.1016/j.cej.2008.07.004

[72] Kralova M, Levchuk I, Kasparek V, Sillanpaa M, Cihlar J. Influence of synthesis conditions on physical properties of lanthanide-doped titania for photocatalytic decomposition of metazachlor. Chinese Journal of Catalysis. 2015;**36**:1679-1685. DOI: 10.1016/S1872-2067(15)60943-3

[73] Leal EGV, Neira JPL, Avella JAP, Pérez E, Meza O. Screening of factors influencing the photocatalytic activity of TiO_2:Ln (Ln = La, Ce, Pr, Nd, Sm, Eu and Gd) in the degradation of dyes. Computational Materials Science. 2015;**107**:48-53. DOI: 10.1016/j.commatsci.2015.05.014

[74] Reszczyńska J, Grzyb T, Wei Z, Klein M, Kowalska E, Ohtani B, et al. Photocatalytic activity and luminescence properties of RE^{3+}–TiO_2 nanocrystals prepared by sol–gel and hydrothermal methods. Applied Catalysis B: Environmental. 2016;**181**:825-837. DOI: 10.1016/j.apcatb.2015.09.001

[75] Li H, Zheng K, Sheng Y, Song Y, Zhang H, Huang J, et al. Facile synthesis and luminescence properties of TiO_2:Eu^{3+} nanobelts. Optics and Laser Technology. 2013;**49**:33-37. DOI: 10.1016/j.optlastec.2012.12.007

[76] Jun D, Buhui L, Jingjing H, Wenlong Z, Hailong P, Jianguo Z. Hydrophilic and photocatalytic performances of lanthanum doped titanium dioxide thin films. Journal of Rare Earths. 2013;**31**(10):992-996. DOI: 10.1016/S1002-0721(13)60019-8

[77] Tang J, Chen X, Liu Y, Gong W, Peng Z, Cai T, et al. Samarium-doped mesoporous TiO_2 nanoparticles with improved photocatalytic performance for elimination of gaseous organic pollutants. Solid State Sciences. 2013;**15**:129-146. DOI: 10.1016/j.solidstatesciences.2012.10.001

[78] Xiao Q, Si Z, Zhang J, Xiao C, Tan X. Photoinduced hydroxyl radical and photocatalytic activity of samarium-doped TiO_2 nanocrystalline. Journal of Hazardous Materials. 2008;**150**:62-67. DOI: 10.1016/j.jhazmat.2007.04.045

[79] Giri PK, Galvagno G, Ferla A, Rimini E, Coffa S, Raineri V. Formation and annealing of defects during high-temperature processing of ion-implanted epitaxial silicon: the role of dopant implants. Materials Science and Engineering. 2000;**B71**:186-191. DOI: 10.1016/S0921-5107(99)00372-4

[80] Mamane H, Horovitz I, Lozzi L, Camillo D, Avisar D. The role of physical and operational parameters in photocatalysis by N-doped TiO_2 sol–gel thin films. Chemical Engineering Journal. 2014;**257**:159-169. DOI: 10.1016/j.cej.2014.07.018

[81] Saharudin KA, Sreekantan S, Lai CW. Fabrication and photocatalysis of nanotubular C-doped TiO_2 arrays: Impact of annealing atmosphere on the degradation efficiency of methyl orange. Materials Science in Semiconductor Processing. 2014;**20**:1-6. DOI: 10.1016/j.mssp.2013.12.019

[82] Acevedo-Peña P, Carrera-Crespo JE, González F, González I. Effect of heat treatment on the crystal phase composition, semiconducting properties and photoelectrocatalytic color removal efficiency of TiO_2 nanotubes arrays. Electrochimica Acta. 2014;**140**:564-571. DOI: 10.1016/j.electacta.2014.06.056 0013-4686

[83] Borlaf M, Caes S, Dewalque J, Colomer MT, Moreno R, Cloots R, et al. Effect of the RE (RE = Eu, Er) doping on the structural and textural properties of mesoporous TiO_2 thin films obtained by evaporation induced self-assembly method. Thin Solid Films. 2014;**558**:140-148. DOI: 10.1016/j.tsf.2014.03.002

[84] Elghniji K, Hentati1 O, Mlaik N, Mahfoudh A, Ksibi M. Photocatalytic degradation of 4-chlorophenol under P-modified TiO_2/UV system: Kinetics, intermediates, phytotoxicity and acute toxicity. Journal of Environmental Sciences. 2012;**24**(3):479-487. DOI: 10.1016/S1001-0742(10)60659-6

[85] Armaković SJ, Grujić-Brojčin M, Šćepanović M, Armaković S, Golubović A, Babić B, et al. Efficiency of La-doped TiO_2 calcined at different temperatures in photocatalytic degradation of β-blockers. Arabian Journal of Chemistry. DOI: 10.1016/j.arabjc.2017.01.001. In press

[86] Đorđević V, Milićević B, Dramićanin MD. Rare earth-doped anatase TiO_2 nanoparticles. In: Janus M, editor. Titanium Dioxide. 1st ed. London: Intech Open; 2017. pp. 25-60. DOI: 10.5772/intechopen.68882

[87] Chen Q, Jiang D, Shi W, Wu D, Xu Y. Visible-light-activated Ce–Si co-doped TiO_2 photocatalyst. Applied Surface Science. 2009;**255**:7918-7924. DOI: 10.1016/j.apsusc.2009.04.167

[88] Cruz D, Arévalo JC, Torres G, Margulis RGB, Ornelas C, Elguézabal AA. TiO_2 doped with Sm^{3+} by sol–gel: Synthesis, characterization and photocatalytic activity of diuron under solar light. Catalysis Today. 2011;**166**:152-158. DOI: 10.1016/j.cattod.2010.08.023

[89] Yang L, Liu P, Li X, Li S. The photo-catalytic activities of neodymium and fluorine doped TiO_2 nanoparticles. Ceramics International. 2012;**38**:4791-4796. DOI: 10.1016/j.ceramint.2012.02.067

Chapter 6

Carbon-/Zeolite-Supported TiO_2 for Sorption/ Photocatalysis Applications in Water Treatment

Florica Manea and Corina Orha

Additional information is available at the end of the chapter

http://dx.doi.org/10.5772/intechopen.80803

Abstract

The role of various carbon forms, i.e., activated carbon and carbon nanotubes/nanofibers as support for TiO_2 in drinking water treatment, is discussed. Also, TiO_2 supported onto zeolite that acts bifunctionally as a sorbent/photocatalyst for drinking water treatment is presented. The main contaminants of natural organic matter (NOM), arsenic species, and nitrogen compounds from drinking water sources by the type of groundwater and surface water can be removed/degraded by sorption/photocatalysis using TiO_2 supported onto carbon and/or zeolite. TiO_2 supported on powdered activated carbon (PAC-TiO_2), granular activated carbon (GAC-TiO_2), and zeolite (Z-TiO_2), namely, *supported* TiO_2, was synthesized through the sol-gel method, and TiO_2 and multiwall carbon nanotubes/carbon nanofibers dispersed within epoxy matrix (CNT-TiO_2-Epoxy, CNF-TiO_2-Epoxy), namely, TiO_2 *composite*, were obtained through the two-roll mill method. Kinetic study results through specific mathematic models allowed to elucidate some mechanistic aspects for sorption and photocatalysis for the application in drinking water. The intercalation of the carbon- and zeolite-supported TiO_2 layers into a filtering system allows to develop a self-cleaning filtering system in drinking water.

Keywords: carbon-supported TiO_2, zeolite-supported TiO_2, sorption, photocatalysis, drinking water treatment

1. Introduction

Water pollution is a concern for the European population, and the quality objectives for the water protection are set through the EU Water Framework Directive. Also, one of the targets of the Clean Water and Sanitation Goal within the Millennium Sustainable Development

Goals of the 2030 Agenda is to achieve universal and equitable access to safe and affordable drinking water for all by 2030 [1].

The challenges in treating drinking water are dependent on the water sources, mainly consisted of surface and groundwater. Problematic substances in the drinking water sources can include organic matter and/or different inorganic contaminants, e.g., nitrite, nitrate, and ammonium.

One of the common and advanced unitary processes used in drinking water technology is sorption designed as activated carbon- or zeolite-based filtering. Due to more hydrophobic nature of the activated carbon, it has been recognized for elimination of a broad range of hazardous materials belong to the organic class from aqueous solutions. Activated carbon is less efficiently applied for removal of simple inorganic metallic ions or small-size anions [2].

Zeolite is also a known filtering material that exhibits a high sorption capacity [3] characterized by a selectivity degree in relation with the type of zeolite based on its specific properties of molecular sieve and ionic exchanger [4–6]. The main disadvantage is represented by the fouling of the filtering material surface.

Photocatalysis process is considered very promising for advanced water treatment based on the oxidation and reduction reaction for in situ generation of highly reactive transitory species (i.e., H_2O_2, OH, O_2, O_3) for mineralization of organic compounds and disinfection by-products [7–8], through the electron-hole pair formation under UV or solar irradiation.

Among various semiconductors that have been investigated in photocatalysis application, titanium dioxide (TiO_2) has attracted much attention due to its physical and chemical stability, negligible toxicity, the resistance to corrosion, redox selectivity, high photostability, and easy preparation [3, 9–10]. By UV irradiation onto TiO_2 surface, under conditions of photon energy (hv) greater than or equal to the bandgap energy of TiO_2, the electron will be photoexcited from the valence band to the empty conduction band leading to an empty unfilled valence band that corresponded to the hole and thus creating the electron-hole pair. The electron-hole pair is involved within various oxidative/reductive reactions including the degradation of organics.

For the drinking water treatment, the main disadvantage of TiO_2-based photocatalysis is given by the necessity of the further separation phase to remove TiO_2 from water, in order to avoid the loss of catalyst particles and introduction of the new pollutant of contamination of TiO_2 in the treated water [11]. A solution to avoid the introduction of a new separation phase includes catalyst fixation onto various supports, e.g., activated carbon [12] and mesoporous clays [13]. The catalyst immobilization on different supports allows getting catalyst composites, which are considered a new generation of catalyst with different properties in relation with those of solely TiO_2. According to the obtaining methods, there are a large variety of composites [14], from which in this chapter will be discussed:

- Composites with TiO_2 coated (supported) on the support of activated carbon by the type of powder and granular and zeolite [15–21]
- Hybrid composites, which comprise a matrix material of polymer (epoxy) in which TiO_2 and the support of carbon nanotubes or nanofibers are dispersed [22–24]

Taking into account that the catalyst represents the key of the performance of the photocatalytic application, the catalyst-based composite should exhibit various effects related to the components and the obtaining methods [14], such as simple supporting effect, stabilizing the microstructure or active components, formation of new compounds which act as active components or stabilizers, having two or more functions, controlling of redox performances, and influencing the kinetics of adsorption/desorption and diffusion of molecules.

In this chapter, two types of composites obtained by two different methods and applied in the drinking water treatment are presented:

- TiO_2 supported on powdered activated carbon (PAC-TiO_2), granular activated carbon (GAC-TiO_2), and zeolite (Z-TiO_2), namely, *supported TiO_2*.
- TiO_2 and multiwall carbon nanotubes and carbon nanofibers dispersed within epoxy matrix (CNT-TiO_2-Epoxy, CNF-TiO_2-Epoxy), namely, *TiO_2 composite*. The carbon and zeolite components were selected taking into account their utility as sorbent for natural organic matter [20–24].

2. TiO_2-based composites obtaining and morphostructural properties

The granular, powdered activated carbon and natural zeolite were functionalized with TiO_2 using titanium tetraisopropoxide (TTIP) as TiO_2 precursors by sol-gel method (**Figure 1**).

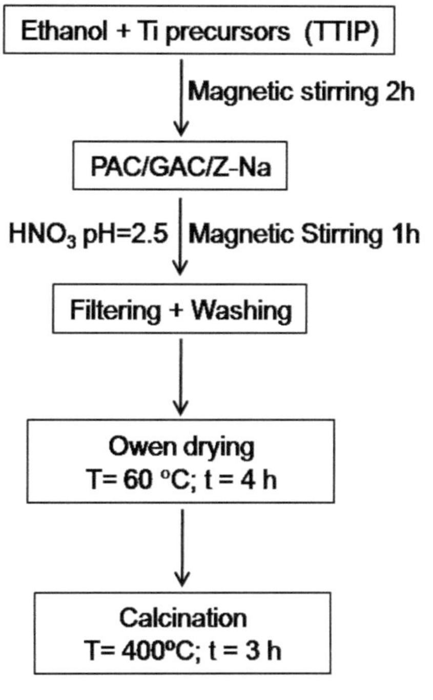

Figure 1. Schematic diagram of obtaining PAC-TiO_2, GAC-TiO_2, and Z-TiO_2.

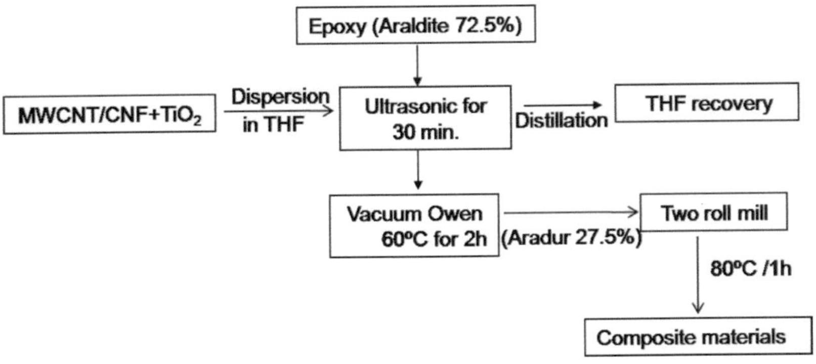

Figure 2. Schematic diagram of obtaining CNT-TiO_2-Epoxy and CNF-TiO_2-Epoxy.

Figure 3. SEM images for (a) PAC-TiO_2, (b) GAC-TiO_2, and (c) Z-TiO_2.

For the synthesis process of the composite materials consisting of carbon nanotubes or carbon nanofibers mixed with TiO_2 particles within epoxy matrix, the two-roll mill technique was applied (**Figure 2**).

http://dx.doi.org/10.5772/intechopen.80803

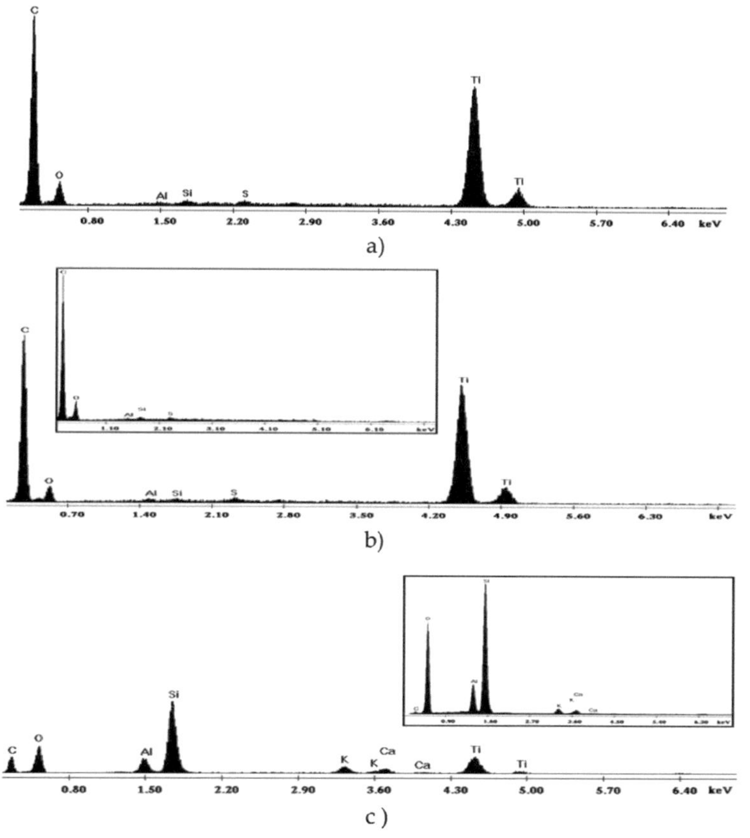

Figure 4. EDX spectra for (a) PAC-TiO_2, (b) GAC-TiO_2, (inset: GAC), and (c) Z-TiO_2 (inset: zeolite).

The morphology of all synthesized materials was observed by a scanning electronic microscope (SEM, Inspect S PANalytical model) coupled with the energy-dispersive X-ray analysis detector (EDX). **Figure 3a–c** presents SEM images for PAC-TiO_2, GAC-TiO_2, and Z-TiO_2, which show the layered structure of activated carbon and a nonuniform distribution and more or less agglomeration of TiO_2 nanoparticles onto the surface. It can be also seen that TiO_2 only adhered to the surface of zeolite without the insight of the inside of the zeolite pores.

The results of semiquantitative elemental analysis of the synthesized material surface were presented by EDX spectra indicating Ti and C presence for PAC-TiO_2 and GAC-TiO_2 and, also, the presence of Ti on the natural zeolite (**Figure 4**). Small amounts of Al, Si, and S were identified within the activated carbon as both powdered and granular as impurities and the presence of K and Ca within the zeolite composition were found.

For composite materials that used epoxy as matrix, it can be noticed that TiO_2 nanoparticles were attached on the nanostructured carbon surface, and, also, a well dispersion of carbon and TiO_2 nanoparticles within epoxy matrix is noticed (**Figure 5a,b**). A slight more uniform distribution of TiO_2 nanoparticles seems to be for CNF in comparison with CNT [22]. The presence of TiO_2 particles in composite materials was confirmed by EDX spectra (**Figure 6**).

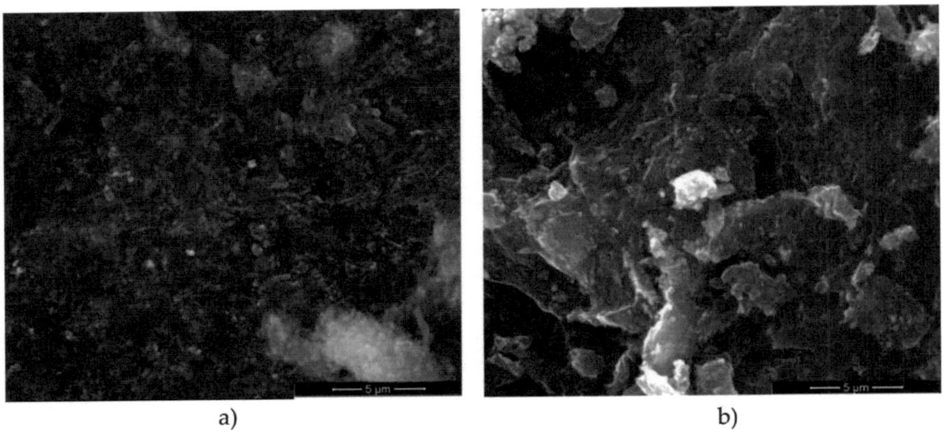

Figure 5. SEM images for (a) CNF-TiO_2-Epoxy and (b) CNT-TiO_2-Epoxy.

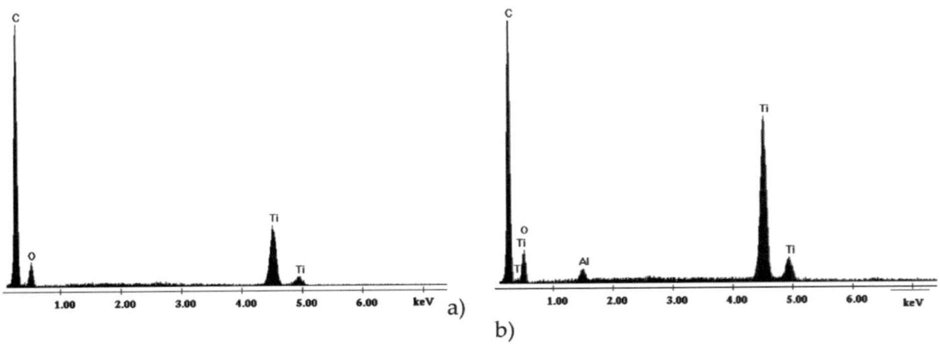

Figure 6. EDX spectra for (a) CNF-TiO_2-epoxy and (b) CNT-TiO_2-epoxy.

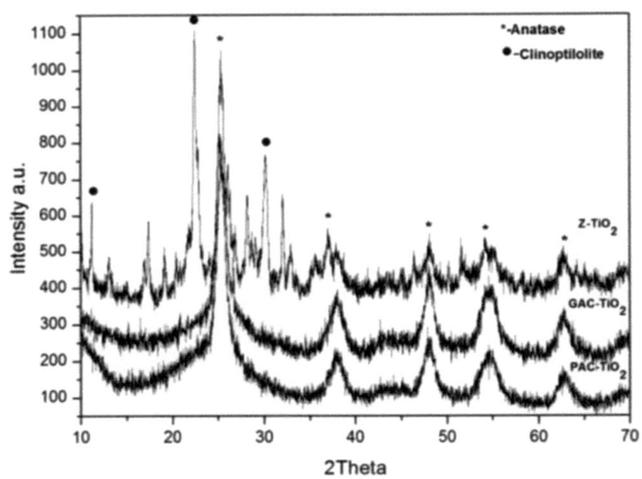

Figure 7. XRD pattern for PAC-TiO_2, GAC-TiO_2, and Z-TiO_2.

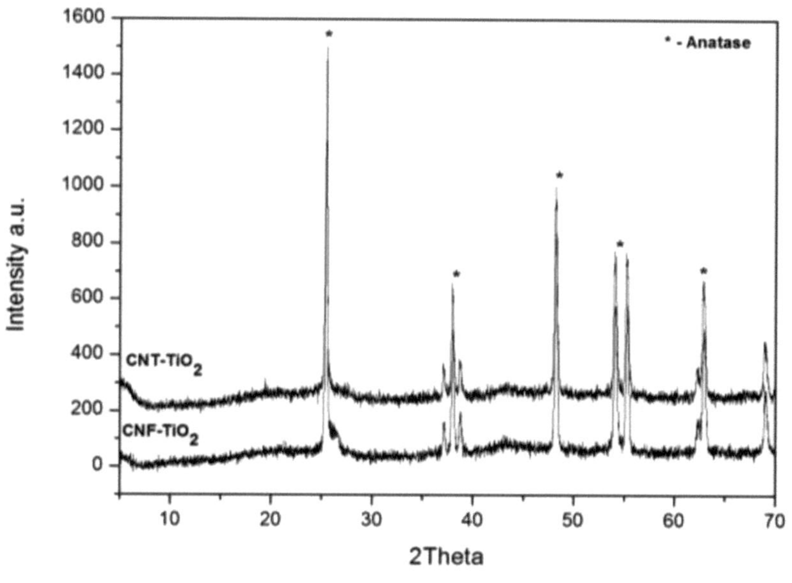

Figure 8. XRD pattern for CNT-TiO_2 and CNF-TiO_2.

X-ray diffraction measurements were carried out to determine the crystal phase composition using a PANalytical X'PertPRO MPD diffractometer. **Figure 7** presents the XRD patterns of GAC-TiO_2, PAC-TiO_2, and Z-TiO_2. The anatase form of TiO_2 is the predominant phase, identified by diffraction lines at 2θ of 25.3, 38.6, 48, 54, and 62.97° [25]. Also, the XRD results revealed that the major component of natural zeolite used in this study is clinoptilolite (2θ ~ 10°; 12°; 22.5°; 30°) [26].

Figure 8 presents the XRD patterns of composite materials based on nanostructured carbon. It is obviously the presence of anatase form of TiO_2 as predominant phase, with the intensity of diffraction lines higher due to the higher TiO_2 loading within the composite composition in comparison with the supported TiO_2.

3. TiO_2-based composite application for sorption and photocatalysis-unitary processes in drinking water treatment

Various types of carbon and zeolite are reported as good sorbents for various types of pollutants from water [2, 15–24]. The easy separation of these materials from water and their good sorption capacity that is important in the first stage of the overall photocatalysis process make them very promising as support for the TiO_2 immobilization.

Humic acids (HAs) represent a main component of natural organic matter that gives the organic loading of drinking water sources, which must be removed or destroyed because of water quality regulations. Also, the presence of HAs before the chlorination step as disinfection could lead to trihalomethane and other toxic by-product generation. The efficiency of *supported TiO_2* materials for the sorption and the photocatalysis of HA from water, expressed as HA removal efficiency, is presented comparatively in **Figure 9**.

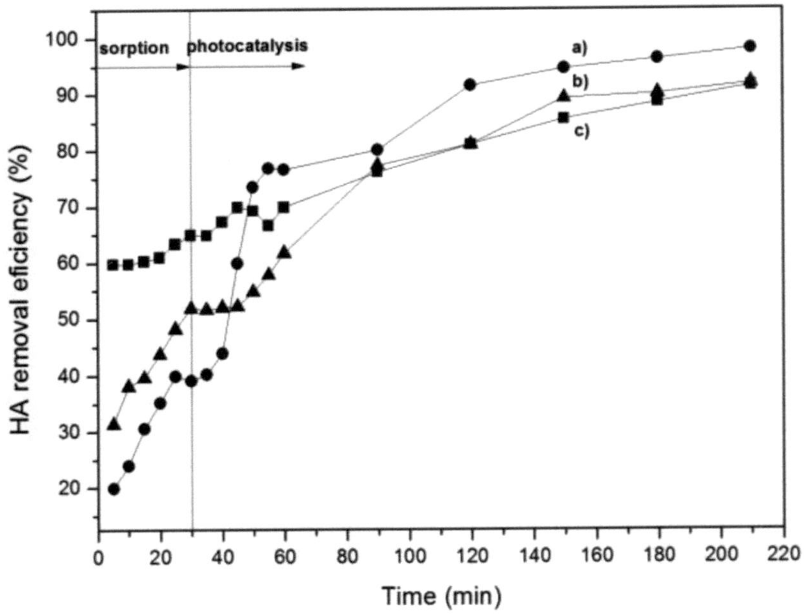

Figure 9. Evolution of HA removal efficiency by sorption and photocatalysis for 50 mg·L^{-1} HA onto (a) GAC-TiO_2, (b) Z-TiO_2, and (c) PAC-TiO_2.

The HA removal efficiency was calculated using the following equation:

$$\text{Removal efficiency} = \frac{C_0 - C_t}{C_0} \times 100\ (\%) \quad (1)$$

where C_0 and C_t are the concentrations of HA in aqueous solution in terms of A_{254} at initial time and at any time t, respectively (mg·L^{-1}).

The sorption was assessed as the preliminary and compulsory step of the overall photocatalysis process taking into consideration also the possibility to use the supports as simple sorbents and only for self-cleaning to apply the photocatalysis. The sorption capacity for HA is better for PAC-TiO_2, while the photocatalytic activity is better for GAC-TiO_2. This should explain the morphological structure and the sizes of the particles. The lower particle size leads to the higher sorption capacity, while for the photocatalysis application the size of support decreased the photocatalytic activity may be due to the finer suspension should hinder the UV irradiation penetration to the TiO_2 surface.The sorption and the photocatalysis capacities of *TiO_2 composite materials* assessed in terms of HA removal efficiency showed the superiority of CNT-TiO_2-Epoxy versus CNF-TiO_2-Epoxy. It can be noticed that a slight photocatalytic activity was found for both materials (**Figure 10**).

For HA removal, the sorption capacity of tested materials increased in order:

$$PAC - TiO_2 > Z - TiO_2 > GAC - TiO_2 > CNT - TiO_2 - Epoxy > CNF - TiO_2 - Epoxy$$

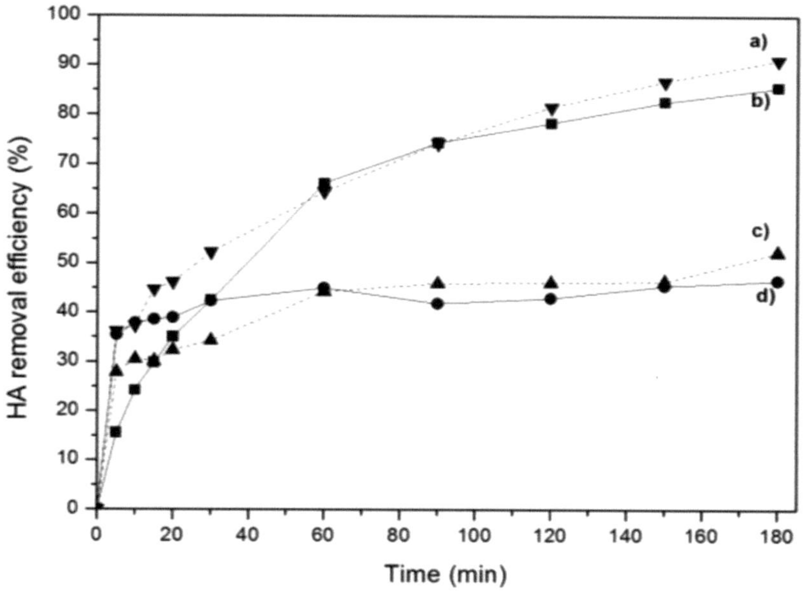

Figure 10. Evolution of removal efficiency of 25 mg·L^{-1} HA using CNT-TiO_2-epoxy by photocatalysis (a), sorption (b), CNF-TiO_2-epoxy by photocatalysis (c), and sorption (d).

The evaluation of photocatalytic activity after the sorption step showed an increasing order:

$$GAC - TiO_2 > Z - TiO_2 > PAC - TiO_2 > CNT - TiO_2 - Epoxy > CNF - TiO_2 - Epoxy$$

These behaviors should be explained by the available sites of the carbon in the sorption process and the available sites of TiO_2 in the photocatalysis process, which are in direct relations with the morphology, size, and the bonding type between TiO_2, carbon, and zeolite support. Also, epoxy resin hindered the available sites, which decreased both sorption capacity and photocatalytic activity of *TiO_2 composite materials*.

In order to determine some mechanism aspects regarding the sorption and photocatalysis processes, several kinetic models must be checked to find the optimum one. Two different kinetic models are used to fit the experimental data, i.e., pseudo-first-order and pseudo-second-order kinetic models [27].

3.1. Pseudo-first-order kinetic model

The pseudo-first-order kinetic adsorption was suggested by Lagergren (1898) for the adsorption of solid/liquid systems. It can be expressed in integrated form as shown in Eq. (2):

$$ln(q_e - q_t) = ln(q_e) - k_1 t \qquad (2)$$

where, k_1 is the rate constant of adsorption (min^{-1}) and q_t is the adsorption loading of dye (mg·g^{-1}) at time t (min).

3.2. Pseudo-second-order kinetic model

The pseudo-second-order kinetics, proposed by Ho and Mckay [28], is expressed in Eq. (3):

$$\frac{t}{q_t} = \frac{1}{k_2 \cdot q_e^2} + \frac{t}{q_e} \quad (3)$$

where, k_2 is the rate constant of the pseudo-second-order adsorption kinetics ($g{\cdot}mg^{-1}{\cdot}min^{-1}$) and q_e is the equilibrium adsorption capacity ($mg{\cdot}g^{-1}$).

Pseudo-first- and pseudo-second-order kinetic models were tested for fitting the sorption and photocatalysis experimental data (**Tables 1** and **2**). The linear plots of *t*/*qt* vs. *t* show that the experimental data agree with the pseudo-second-order kinetic model for the HA adsorption. The calculated *qe* values agree very well with the experimental data, and the correlation coefficients for the pseudo-second-order kinetic model are higher than 0.90 in almost all cases. These indicate that the adsorption of HA from water onto materials obeys the pseudo-second-order kinetic model that predicts an exponential decay of concentrations as a function of time. The pseudo-first-order kinetic model was not appropriate for fitting the sorption experimental data, while the photocatalysis experimental data were fitted well with both kinetic models with the correlation coefficient

Catalyst	Parameters	Adsorption	Photocatalysis
		HA concentration (50 mg·L^{-1})	
PAC-TiO_2	q_e(mg*g^{-1})	38.34	44.25
	k_2(g·mg^{-1}·min^{-1})	0.0067	0.0031
	R^2	0.915	0.996
GAC-TiO_2	q_e(mg*g^{-1})	25.59	49.78
	k_2(g·mg–1·min^{-1})	0.00381	0.0170
	R^2	0.966	0.998
Z-TiO_2	q_e(mg*g^{-1})	28.65	42.32
	k_2(g·mg^{-1}·min^{-1})	0.0059	0.0025
	R^2	0.976	0.985
		HA concentration (25 mg·L^{-1})	
CNT-TiO_2-Epoxy	q_e(mg*g^{-1})	20.68	19.98
	k_2(g·mg^{-1}·min^{-1})	0.0024	0.0015
	R^2	0.986	0.996
CNF-TiO_2-Epoxy	q_e(mg*g^{-1})	13.06	12.37
	k_2(g·mg^{-1}·min^{-1})	0.0045	0.0001
	R^2	0.996	0.943

Table 1. Pseudo-second-order kinetic parameters for HA sorption and photocatalysis.

Catalyst	Parameters	Value
	HA concentration (50 mg·L⁻¹)	
PAC-TiO_2	q_e(mg*g^{-1})	9.95
	k_1(min^{-1})	0.0073
	R^2	0.955
GAC-TiO_2	q_e(mg*g^{-1})	1.64
	k_1(min^{-1})	0.0494
	R^2	0.913
Z-TiO_2	q_e(mg*g^{-1})	2.92
	k_1(min^{-1})	0.0142
	R^2	0.943
	HA concentration (25 mg·L⁻¹)	
CNT-TiO_2-Epoxy	q_e(mg*g^{-1})	0.33
	k_1(min^{-1})	0.085
	R^2	0.995
CNF-TiO_2-Epoxy	q_e(mg*g^{-1})	1.66
	k_1(min^{-1})	0.0019
	R^2	0.956

Table 2. Pseudo-first-order kinetic parameters for HA photocatalysis.

higher than 0.90. However, there is a significant difference between calculated and experimental qe using the pseudo-first-order kinetic model that limits its interpretation (**Table 2**), and, thus, it means that the pseudo-second-order kinetic model is most appropriate. Besides the kinetic parameter, the pseudo-second-order kinetic model informed about the significance of the sorption step within the overall photocatalysis process. It is well known that the reaction rate and kinetics are influenced by many experimental parameters, *e.g.*, HA concentration, catalyst dose, and pH [22].

Based on the results presented in **Table 1**, it can be seen that the PAC-TiO_2 exhibited higher sorption capacity and the best kinetics for HA removal and the GAC-TiO_2 exhibited the best photocatalytic kinetics, which are in accordance with the efficiency results. The worse results were achieved for *TiO_2 composite* materials, which can be explained by the presence of epoxy matrix that reduced the active sites for the sorption and the photocatalysis. Also, the main difference was found between CNT-TiO_2-Epoxy and CNF-TiO_2-Epoxy due to the morphology and sizes of the nanostructured carbon component, better for CNT in comparison with CNF. For these types of material, the photocatalytic activity was better also for CNT-TiO_2-Epoxy.

Arsenic is a common contaminant in drinking water supplies especially for groundwater sources. Inorganic arsenic speciation in water consists of arsenite (AsIII) and arsenate (AsV), but As(III) is more problematic because of its high toxicity and the difficulty to be removed from

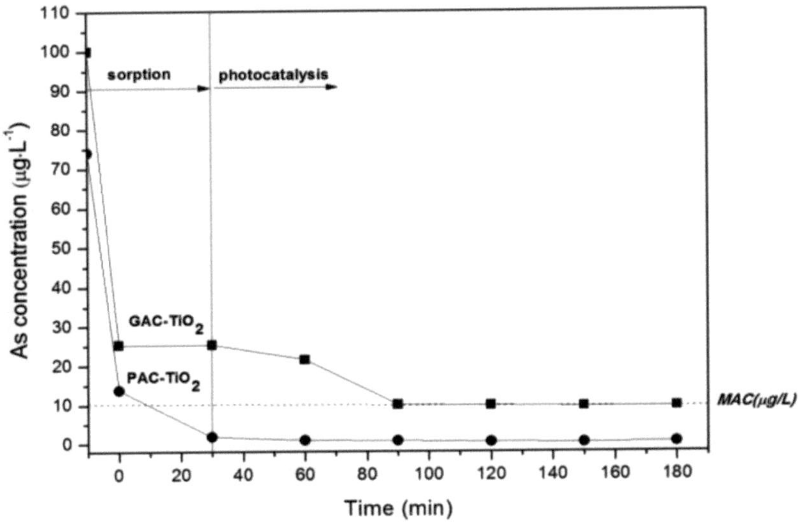

Figure 11. Time evolution of As(III) concentration during sorption and photocatalysis using PAC-TiO_2 and GAC-TiO_2.

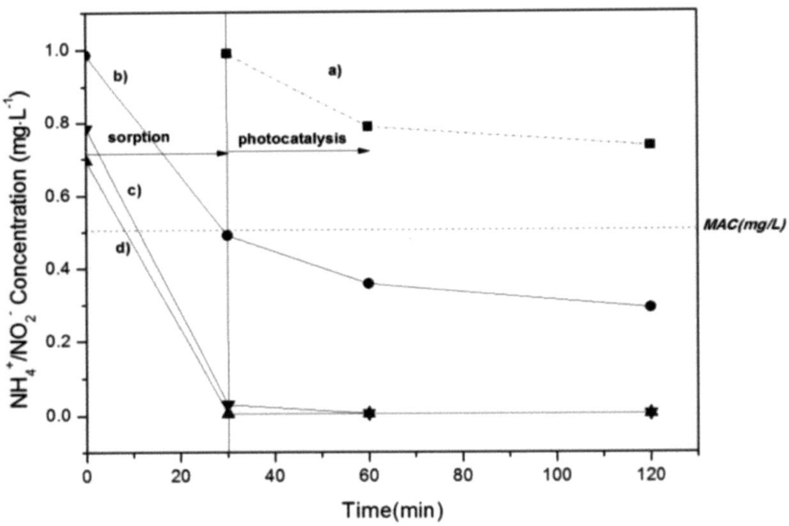

Figure 12. Time evolution of NO_2^- concentration during photolysis (a), photocatalysis using Z-TiO_2 (b), of NH_4^+ concentration during photocatalysis using Z-TiO_2 (c), and natural zeolite (d).

water [29]. The comparative results regarding As(III) removal in the presence of HA showed a good efficiency for both sorption and photocatalysis using *supported* TiO_2 material (**Figure 11**).

Another common problem of groundwater source used for drinking water supply is given by the presence of ammonium, nitrite, and nitrate. Nitrate could not be removed from water on these tested materials, while ammonium and nitrate were removed from Z-TiO_2 (**Figure 12**).

http://dx.doi.org/10.5772/intechopen.80803

The results showed that ammonium is removed by sorption on the zeolite based on ionic exchange, which is proven by comparison with natural zeolite, while the nitrite removal is based on its oxidation process during UV irradiation, which is enhanced by the presence of the Z-TiO_2 photocatalyst.

Since activated carbon is a versatile sorbent class for the retention of a wide spectrum of organic compounds and arsenic(III) and the zeolite exhibited an adsorption/ion-exchange capacity for ammonium, nitrite, and heavy metals, their common presence as mixed materials broadens the spectrum of contaminants that can be removed from the water.

4. Conclusions

In this chapter, two types of composites obtained by two different methods and applied in the drinking water treatment were studied:

a. TiO_2 supported on powdered activated carbon (PAC-TiO_2), granular activated carbon (GAC-TiO_2), and zeolite (Z-TiO_2), namely, *supported TiO_2*, obtained through sol-gel method.

b. TiO_2 and multiwall carbon nanotubes and carbon nanofibers dispersed within epoxy matrix (CNT-TiO_2-Epoxy, CNF-TiO_2-Epoxy), namely, *TiO_2 composite*, obtained through the two-roll mill method.

The sorption and photocatalysis studies showed that for HA removal, the sorption capacity of tested materials increased in order, PAC-TiO_2 > Z-TiO_2 > GAC-TiO_2 > CNT-TiO_2-Epoxy>CNF-TiO_2-Epoxy, and for the photocatalytic activity, GAC-TiO_2 > Z-TiO_2 > PAC-TiO_2 > CNT-TiO_2-Epoxy> CNF-TiO_2-Epoxy.

The removal of pollutants/impurities dissolved in water is based, on the one hand, on the sorption capacity of activated carbon and/or zeolite, which exhibited selectivity for certain impurities dissolved under the lamp-off conditions. On the other hand, but taking into account the sorption phase, by starting the lamp, TiO_2 on the active carbon surface and the zeolite acted by photocatalytic activity generating oxidation and reduction processes that cause the degradation and mineralization of the dissolved organic compounds and the transformation of the inorganic contaminants into the compounds that can be retained on the surface material or compounds that do not affect the quality of the water. Also, the oxidation and reduction processes under irradiation conditions allow the destruction of the contaminants and, implicitly, the cleaning of the sorbent surface. A mixture of zeolite and activated carbon as support for TiO_2 could exhibit bifunctionality, depending on the water features. Thus, for low loading of water with contaminants that can be adsorbed on the filter material, the lamp is not switched on during system operation but only in the washing/regeneration stage, and for a water loaded with contaminants that cannot be adsorbed on the filter material, the system works with the lamp on, ensuring water decontamination and self-cleaning during operation.

Acknowledgements

This work was supported by the Romanian National Research Programs PN-III-P2-2.1-PED-2016, PED 69/2017.

Author details

Florica Manea[1]* and Corina Orha[2]

*Address all correspondence to: florica.manea@upt.ro

1 Faculty of Industrial Chemistry and Environmental Engineering, Politehnica University of Timisoara, Timisoara, Romania

2 Condensed Matter Department, National Institute for Research and Development in Electrochemistry and Condensed Matter, Timisoara, Romania

References

[1] European Commission Environment, Water Framework Directive. Available from: http://ec.europa.eu/environment/water/water-framework/index_en.html

[2] Orha C, Lazau C, Pode R, Manea F. Simultaneous removal of humic acid and arsenic (III) from drinking water using TiO_2-powdered activated carbon. Journal of Environmental Protection and Ecology. 2018;**19**:39-47

[3] Nan Chong M, Jin B, Chow WK, Saint C. Recent developments in photocatalytic water treatment technology: A review. Water Research. 2010;**44**:2997-3027. DOI: 10.1016/j.watres.2010.02.039

[4] Wang S, Peng Y. Natural zeolites as effective adsorbents in water and wastewater treatment. Chemical Engineering Journal. 2010;**156**:11-24. DOI: 10.1016/j.cej.2009.10.029

[5] Paliulis D. Removal of formaldehyde from synthetic wastewater using natural and modified zeolites. Polish Journal of Environmental Studies. 2016;**25**:251-257. DOI: 10.15244/pjoes/60727

[6] Capasso S, Salvestrini S, Coppola E, Buondonno A, Colella C. Site-energy distribution analysis of humic acids adsorption by zeolitic tuff. Journal of Porous Materials. 2007;**14**:363-368. DOI: 10.1007/s10934-006-9074-2

[7] Esplugas S, Gimenez J, Conteras S, Pascual E, Rodrıguez M. Comparison of different advanced oxidation processes for phenol degradation. Water Research. 2002;**36**:1034-1042. DOI: 10.1016/S0043-1354(01)00301-3

[8] Pera-Titus M, Garcıa-Molina V, Banos MA, Gimenez J, Esplugas S. Degradation of chlorophenols by means of advanced oxidation processes: A general review. Applied Catalysis B: Environmental. 2004;**47**:219-256. DOI: 10.1016/j.apcatb.2003.09.010

[9] Gaya UI, Abdullah AH. Heterogeneous photocatalytic degradation of organic contaminants over titanium dioxide: A review of fundamentals, progress and problems. Journal of Photochemistry and Photobiology, C: Photochemistry Reviews. 2008;**9**:1-12. DOI: 10.1016/j.jphotochemrev.2007.12.003

[10] Fujishima A, Rao TN, Tryk DA. Titanium dioxide photocatalysis. Journal of Photochemistry and Photobiology, C: Photochemistry Reviews. 2000;**1**:1-21. DOI: 10.1016/S1389-5567(00)00002-2

[11] Yang GCC, Li CJ. Electrofiltration of silica nanoparticle containing wastewater using tubular ceramic membranes. Separation and Purification Technology. 2007;**58**:159-165. DOI: 10.1016/j.seppur.2007.07.019

[12] Lee DK, Kim SC, Cho IC, Kim SJ, Kim SW. Photocatalytic oxidation of microcystin-LR in a fluidized bed reactor having TiO_2-coated activated carbon. Separation and Purification Technology. 2004;**34**:59-66. DOI: 10.1016/S1383-5866(03)00175-8

[13] Chong MN, Vimonses V, Lei S, Jin B, Chow C, Saint C. Synthesis and characterisation of novel titania impregnated kaolinite nano-photocatalyst. Microporous and Mesoporous Materials. 2009;**117**:233-242. DOI: 10.1016/j.micromeso.2008.06.039

[14] Xie Z, Liu Z, Wang Y, Yang Q, Xu LDW. An overview of recent development in composite catalysts from porous materials for various reactions and processes. International Journal of Molecular Sciences. 2010;**11**:2152-2187. DOI: 10.3390/ijms11052152

[15] Orha C, Lazau C, Manea F, Ursu D. Effect of carbon nanofiber onto TiO_2-modified powder/granular activated carbon for advanced water treatment. In: Proceedings of the 23rd International Symposium on Analytical and Environmental Problems; 9-10 October 2017; University of Szeged, Department of Inorganic and Analytical Chemistry. pp. 244-248

[16] Orha C, Pode R, Manea F, Lazau C, Bandas C. Titanium dioxide-modified activated carbon for advanced drinking water treatment. Process Safety and Environment Protection. 2017;**108**:26-33. DOI: 10.1016/j.psep.2016.07.013

[17] Lazau C, Ratiu C, Orha C, Pode R, Manea F. Photocatalytic activity of undoped and Ag-doped TiO_2-supported zeolite for humic acid degradation and mineralization. Materials Research Bulletin. 2011;**46**:1916-1921. DOI: 10.1016/j.materresbull.2011.07.026

[18] Orha C, Pop A, Lazau C, Jakab A, Pode R, Manea F. Carbon-zeolite composite materials in advanced drinking water treatment. In: Borchers U, Gray J, Thompson KC, editors. Water Contamination Emergencies: Managing the Threats. 2013. pp. 346-355. DOI: 10.1039/9781849737890-00346

[19] Ratiu C, Manea F, Lazau C, Orha C. Zeolite-supported TiO_2 based photocatalysis-assisted electrochemical degradation of p-aminophenol from water. Chemical Papers. 2011;**65**:289-298. DOI: 10.2478/s11696-011-0009-2

[20] Ilinoiu EC, Pode R, Manea F, Colar LA, Jakab A, Orha C, et al. Photocatalytic activity of a nitrogen doped TiO_2 modified zeolite in the degradation of reactive yellow 125 azo dye. Journal of the Taiwan Institute of Chemical Engineers. 2013;**44**:270-278. DOI: 10.1016/j.jtice.2012.09.006

[21] Orha C, Lazau C, Ursu D, Manea F. Effect of TiO_2 loading on powder-activated carbon in advanced drinking-water treatment. WIT Transactions on Ecology and the Environment. 2017;**216**:203-211. DOI: 10.2495/WS170191

[22] Orha C, Manea F, Pop A, Bandas C, Lazau C. TiO_2 nanostructured carbon composite sorbent/photocatalyst for humic acid removal from water. Desalination and Water Treatment. 2016;**57**:14178-14187. DOI: 10.1080/19443994.2015.1062433

[23] Orha C, Pop A, Tiponut V, Manea F. Composite material based on silver-doped zeolite and multi-wall carbon nanotubes for humic acid removal. Environmental Engineering and Management Journal. 2013;**12**:917-921

[24] Jakab A, Pode R, Pop A, Schoonman J, Orha C, Manea F. TiO_2-modified zeolite-carbon nanotubes composite electrode for photoelectrodegradation of pentachlorophenol from water under UV irradiation. WIT Transactions on Ecology and the Environment. 2017; **216**:133-142. DOI: 10.2495/WS170191

[25] Neren Ökte A, Sayınsöz E. Characterization and photocatalytic activity of TiO_2 supported septiolite catalysts. Separation and Purification Technology. 2008;**62**:535-543. DOI: 10.1016/j.seppur.2008.03.011

[26] Ratiu C, Lazau C, Sfirloaga P, Orha C, Sonea D, Novaconi S, et al. Decontaminate effect of the functionalized materials with undoped and doped (Ag) TiO_2 nanocrystals. Environmental Engineering and Management Journal. 2009;**8**:237-242

[27] Lv L, He J, Wei M, Evans DG, Duan X. Uptake of chloride ions from aqueous solutions by calcined layered double hydroxides: Equilibrium and kinetic studies. Water Research. 2006;**40**:735-743. DOI: 10.1016/j.watres.2005.11.043

[28] Ho YS, McKay G. Pseudo-second order model for sorption processes. Process Biochemistry. 1999;**34**:451-465. DOI: 10.1016/S0032-9592(98)00112-5

[29] Mohan D, Pittman CU Jr. Arsenic removal from water/wastewater using adsorbents-a critical review. Journal of Hazardous Materials. 2007;**142**:1-53. DOI: 10.1016/j.jhazmat.2007.01.006

Chapter 7

Evaluation of the Role of Hydroxyapatite in TiO_2/Hydroxyapatite Photocatalytic Materials

Linh Nguyen Thi Truc, Seungbum Hong and Kwangsoo No

Additional information is available at the end of the chapter

http://dx.doi.org/10.5772/intechopen.81092

Abstract

The TiO_2/hydroxyapatite (HAp) composite has attracted much attention as a photocatalyst for pollution treatment in water or air because this composite can improve the properties of pure TiO_2 including a low efficiency, narrow light response range, low adsorption capacity for hydrophobic contaminants, and difficult recovery of TiO_2 particles after using in Aquarius environment. To obtain the best composite containing the two components including TiO_2 and HAp, the role of HAp in TiO_2/hydroxyapatite photocatalytic material should be analyzed and evaluated. This chapter will significantly present a review of the role of HAp in the TiO_2/hydroxyapatite composite including the adsorption ability of contaminations and the promoted impacts of HAp component.

Keywords: hydroxyapatite, TiO_2, photocatalytic, absorption, bandgap

1. Introduction

Hydroxyapatite (HAp, $Ca_{10}(PO_4)_6(OH)_2$) is one of the most usual forms of calcium phosphate and has the similar chemical composition to the mineral phase of bone tissues. Thus, HAp has attracted the interest of the scientific community in the medicine field, material science, and tissue engineering areas in many years. However, due to the special characteristics of this material, HAp is also studied for various applications including fluorescent lamps, fuel cells, and adsorption of harmful metals, as well as catalysts [1]. TiO_2 material was concerned almost 48 years ago. For the first time in 1969, the possibility of solar photoelectrolysis of TiO_2 was demonstrated; the powdered TiO_2 was studied in the 1980s and TiO_2 film photocatalysis in the 1990s [2]. TiO_2 photocatalysis (in particles or films) has gained much attention because the material has high stability, low cost, and nontoxicity and can be easily fabricated by many processes including precipitation, hydrothermal, sol-gel, plasma, etc. The research on TiO_2/hydroxyapatite photocatalytic materials may derive from researches on TiO_2/hydroxyapatite composites in medical industry. To achieve biocompatibility, osteoconduction, and osseointegration, the surface of titanium or its alloy, which is used as a permanent implant material, must be modified by developing hydroxyapatite (HAp, $Ca_{10}(PO_4)_6(OH)_2$) coating on the surface [3–9]. By inserting TiO_2 inner layer between HAp coating and Ti substrate, the adhesion strength of the coating and the substrate increases [10]. The addition of TiO_2 inner layer is expected to reduce a thermal expansion mismatch of the layers, to improve the bonding strength between the HAp layer and Ti substrate, to prevent the corrosion of the Ti substrate, as well as to obtain an abundance of surface hydroxyl and superoxide radical groups, sequentially, to achieve a surface free of cracks and a high adhesion of the modified surface to the substrate [7, 11–13]. The reactions of photocatalysis occur on the surface of TiO_2; thus, both the surface properties and the mass transfer of the pollutant and degraded products onto the substrate affect to its photocatalytic activity. HAp, which is known as a material having a large surface area and high adsorption ability, may play the role of a transparent or semi-transparent to allow UV and visible radiation to pass through it. In fact, there are numerous materials having larger surface area and higher adsorption ability such as silica gel, zeolite, activated carbon, etc. However, besides these general properties, HAp material was investigated as a support for photocatalysis due to the generation of active superoxide anion radicals ($O_2^{\cdot -}$) under UV irradiation [14]; thus, the TiO_2/HAp photocatalysts have gained the interest of many scientists. In this chapter, the role of hydroxyapatite in TiO_2/HAp photocatalytic materials will be analyzed and evaluated to supply a clear view of the composites.

2. Hydroxyapatite plays the role of adsorption material in the photocatalytic TiO_2/HAp composites

Adsorption of biological, organic-chemical molecules on the HAp surface is generally influenced by its physicochemical properties including crystallite size, pore structure, morphology of particles, or coatings [15] which directly depend on the synthesis methods. The synthesis methods of HAp particles include solid-state reaction; sol-gel, plasma, and hydrothermal technique; layer hydrolysis of other calcium phosphate salts; etc. [16–18], while those of HAp coating include sol-gel, chemical vapor deposition, pulsed laser deposition, RF magnetron sputtering, spray pyrolysis, etc. [7, 19, 20]. The choice of a specific method depends on the purpose of research which is synthesis, characteristics, or application of pure HAp or those of TiO_2/HAp composites.

The adsorption property of HAp in the photocatalytic TiO_2/HAp composites has been reported in many researches. Nonami et al. [21] soaked TiO_2 powder in a simulated physiological solution containing phosphate ions for periods of about 1 h at 37°C. The apatite film with a thickness of approximately 0.7 μm has formed on the approximately 0.3 μm thick TiO_2 layer. The crystals have a plate-like shape, measuring approximately 0.1–0.5 μm in length and

2–10 nm in thickness. This composite material may adsorb contaminants without exposure to light, and the contaminants are decomposed by TiO_2 photocatalyst on exposure to light. As a result, the photocatalytic TiO_2/HAp composites can be used to pure the air or play the role of the antimicrobial and antifungal coating with HAp absorber.

Other literatures show that the photocatalytic TiO_2/HAp materials not only absorb organic contaminants such as viruses, bacteria, etc. but also decompose these compounds by the photocatalytic process. Hirakura et al. [22] presented that lysozyme (LSZ) and bovine serum albumin (BSA) can be monomolecularly absorbed on HAp by using two types of fibrous crystals elongated in the c-axis. HAp-nanostructured crystals doping TiO_2 anatase can selectively remove of the specific proteins by the absorbing and decomposing under UV irradiation. Thus, the photocatalytic activity for the decomposition of proteins could be controlled with the adsorption on the surface of the nanostructured HAp crystals. The adsorption of BSA on TiO_2/HAp composite was also presented in the publication of Katayama et al. [23]. The research explained that the adsorption of the acidic protein BSA occurred at Ca^{2+} sites of the HAp component which contained a large number of pores supporting to the physical adsorption.

Ryu et al. [24] fabricated the TiO_2-β-TCP nanocomposite photocatalytic thin films by aerosol deposition. The aerosol-deposition films almost fully covered the substrate (glass) and are not porous but extremely rough microstructure. The deposited films maintained good adhesion with the substrate, and the film's pencil hardness was over 9H. The aerosol-deposition TiO_2-40 wt% β-TCP composite film has two different phases distinctly: white regions of β-TCP and dark regions of TiO_2, 5–50 nm-sized TiO_2 nanocrystallites with nanoscaled β-TCP crystallites formed by the collision of the accelerated particles with high kinetic energy during deposition. The films consisting of nano-sized TiO_2 photocatalytic crystallites with a dispersion of β-TCP adsorbent crystalline phase show the high photocatalytic activity under both UV and dark conditions due to adsorption effect of β-TCP.

The adsorption role of the HAp layer in Ag-TiO_2/HAp/Al_2O_3 bioceramic composite membrane was also evaluated in the report of Ma et al. [25]. This composite membrane, which was fabricated by a facile two-step approach, involves sol–gel process and calcination was a microporous membrane structure with average 0.8 m pore size, which comprised of Ag-TiO_2/HAp composite layer with a thickness of 10 m overlaid on α-Al_2O_3 disk support. HAp component acted as a highly efficient bacterial adsorbent, while Ag-TiO_2 provided powerful photocatalytic attack toward *E. coli* strains.

Xie et al. [26] fabricated TiO_2/HAp composite with mosaic structure via a facile route without any structure-directing agent. The result proved the increased photocatalytic activity of the composite results from the combination of adsorption capacity of HAp and the high photocatalytic activity of TiO_2.

Other reports also concerned to the adsorption properties of HAp in the HAp/TiO_2 composites, especially in atmospheric environment [27–29]. Komazaki et al. [27] collected NO(x) by an annular diffusion scrubber coated with a mixture of TiO_2 and HAp. The research shows that HAp plays the role of adsorption material, while TiO_2 produces reactive oxygen species under ultraviolet light (UV) illumination such as super oxide (O_2^-), hydroxyl radical (OH°), and peroxyhydroxyl radical (HO_2°), by which nitric oxide is oxidized to nitrogen dioxide and is further oxidized to nitric acid.

Ozeki et al. [28] fabricated TiO_2/HAp thin film by sputtering on glass using a radio-frequency magnetron. The film has a higher decomposition rate of formaldehyde gas than either the TiO_2 or the HAp film alone. However, in the bacterial survival test, survival of cells on the TiO_2/HAp thin film is higher than that on the TiO_2 film, which indicates that the TiO_2/HAp thin film has a lower bactericidal effect than the pure TiO_2 film.

Liu et al. [29] synthesized HAp-modified N-TiO_2 by a facile wet-chemical method and evaluated its photocatalytic activity by the decomposition of gaseous acetone under visible-light irradiation. The results demonstrated that 10%-HAp-N-TiO_2 sample shows the best photocatalytic activity and the remarkable photocatalytic activity may arise from the synergism between adsorption on HAp and photoactivity by TiO_2 which generated oxygen-reactive species diffusing and reacting with the molecules located on the HAp.

Masato Wakamura et al. have many publications on the field [30–32]. The authors evaluated the adsorption behaviors of proteins onto photocatalytic Ti^{4+}-doped calcium hydroxyapatite (TiHAp) particles which was synthesized by the coprecipitation method using $Ca(NO_3)_2$, $Ti(SO_4)_2$, H_3PO_4, and NH_4OH original materials [31]. The result showed that all the adsorption isotherms of bovine serum albumin (BSA), myoglobin (MGB), and lysozyme (LSZ) from 1×10^{-4} mol/dm^3 KCl solution were Langmuir type. Generally, the saturated amounts of adsorbed protein value on the TiHAp were much higher than CaHAp.

3. The promoted impact of hydroxyapatite in the photocatalytic TiO_2/HAp composites

Beside the role as an adsorption material in the photocatalytic TiO_2/HAp composites, some recent researches have investigated that hydroxyapatite promotes photocatalytic degradation because it can play the role of a support for photocatalysts. The activity of HAp is caused by the generation of active superoxide anion radicals ($O_2^{\cdot -}$) due to a change in the electronic state of the surface PO_4^{3-} group under UV irradiation [14].

Hu et al. [33] proved the promotion role of HAp to TiO_2 in photocatalytic degradation by using experimental observations and kinetic modeling. The derived kinetic parameters including reaction rate constant, Langmuir adsorption constant, apparent activation energy, etc. confirm that the activity of TiO_2/HAp composite is more effective than that of TiO_2. A negative effect of HAp on the photo absorption ability of the TiO_2/HAp composite is characterized by UV-vis reflectance spectra. The existence of superior chemisorption between HAp and the organic molecules leads to a better performance of TiO_2/HAp for photocatalytic degradation.

Mohamed et al. [34] determined photocatalytic activities of Pd-TiO_2-hydroxyapatite nanoparticles, which are synthesized by a template-ultrasonic-assisted method, by photocatalytic removal of cyanide under visible-light irradiation (**Scheme 1**). **Scheme 1** shows that the

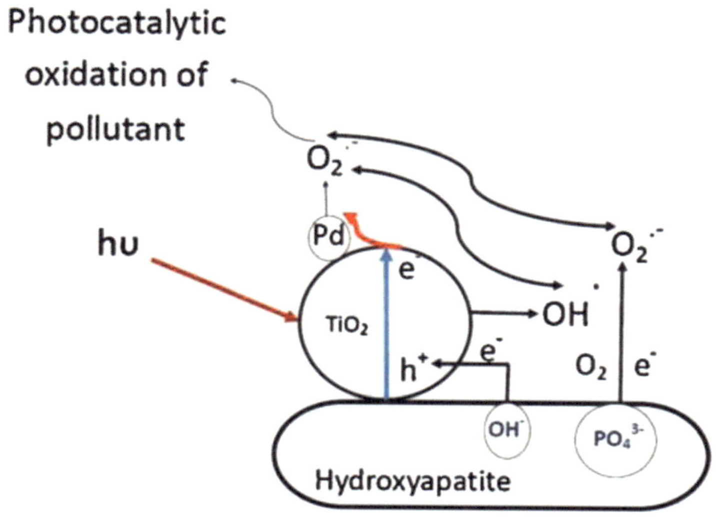

Scheme 1. Photocatalytic mechanism of cyanide degradation by Pd-TiO_2-HAp [34].

vacancies formed on the surface of the excited PO_4^{3-} group in visible illumination will lead to the formation of $O_2^{\cdot-}$ and will attack the surrounding organic molecules adsorbed on HAp.

The research [34] investigated that the amount of hydroxyl radials formed in the Pd-TiO_2-hydroxyapatite nanoparticle was higher than those of Pd-TiO_2 and TiO_2 nanoparticles. The absorption of radiation in the visible-light region, the small recombination rate of the electron-hole pair, as well as the high surface area of HAp material are the promoted impacts of hydroxyapatite in the photocatalytic TiO_2/HAp composites.

The recombination rate of the electron-hole pair is also concerned in another publication [35]. Zhang et al. [35] demonstrated the increased photocatalytic performance of TiO_2 nanoparticles which was supported on electrically polarized HAp films. The separation of photogenerated electrons and holes in TiO_2 nanoparticles is promoted by the internal polarization of the HAp support, and consequently, the recombination of charge carriers is mitigated. It can be concluded that the materials with large internal polarization can be used in strategies for enhancing quantum efficiency of photocatalysts.

The surface area of HAp material was concerned in the research of Kobayashi et al. [36]. The TiO_2/HAp photocatalytic coating was deposited by gas tunnel-type plasma spraying using powders with nano-sized grains. The photocatalytic reaction occurs on powder surface; as the grain size becomes smaller, the specific surface area becomes larger, and the degradation efficiency of organic pollutants also increases.

The promoted impact of hydroxyapatite in the photocatalytic TiO_2/HAp composites was presented in the report of Liu et al. [37]. The photocatalytic activities of Ag-TiO_2-HAp powders synthesized by a facile wet-chemical strategy were evaluated by photocatalytic oxidation decomposition of acetone in air under visible-light illumination. The high photocatalytic activity of the Ag-TiO_2-HAP hybrids could be attributed to its strong absorption in the visible-light region, low recombination rate of the electron-hole pair, and large BET-specific surface area.

Aramendia et al. [38] studied on the TiO_2/natural phosphate material synthesized by sol-gel process and tested the photocatalytic activity in the photo-oxidation process of propan-2-ol. The results show that the presence of natural phosphate makes a retardation of TiO_2 crystallization and creates a certain interaction with titanium.

Nishikawa et al. [39] indicated that the difference in photocatalytic activity of the materials containing HAp and without HAp may be originated from the properties of the electronic state that forms the valence band of HAp-containing PO_4 group.

Li et al. [40] synthesized Ti-substituted hydroxyapatite (TiHAp) by the coprecipitation method. The adsorption and photocatalytic degradation of bisphenol A (BPA, an environmental endocrine disrupting chemical) over TiHAp and P25 TiO_2 photocatalysts were studied using liquid chromatography-mass spectrometry. The results indicated that the adsorption of BPA on TiHAp and TiO_2 obeyed the Langmuir adsorption equation, and the adsorption capacity and photocatalytic degradation activity of BPA of TiHAp material were much higher than those of TiO_2. To explain for the results, the authors presented that the zeta potentials of TiHAp, and TiO_2 did not show significant difference when the pH was about seven, suggesting that the surface charge was not the reason for the different adsorption capacities of the particles. In addition, the specific surface area and average pore diameter of TiHAp and TiO_2 were comparable, so these would not lead to the different adsorption capacities either. In fact, TiHAp material is produced from a substitution of some Ca sites in HAp by Ti, which resulted in multiple Ti-OH groups on the TiHAp surface. Large amounts of phosphates and hydroxyls in the crystal lattice of TiHAp can adsorb the hydroxyls of BPA by hydrogen bonding.

The bandgap of TiO_2/HAp materials were mentioned in some publications because it decides the energy separation between valence and conduction bands, the quantum effect, as well as the effect of visible-light utilization, etc. The bandgap of HAp was reported in [41] to be 3.95 eV by photoluminescence measurement; meanwhile, in other reports, the bandgap of HAp was calculated to be around 4.51–5.4 eV [42, 43]. The bandgap of TiO_2/HAp composites calculated by UV-vis diffuse reflectance spectra was between 3.06 and 3.08 eV while that of pure TiO_2 was broader (3.12 eV) [44]. However, the bandgap of Ti-substituted hydroxyapatite evaluated by both experimental and theoretical methods was 3.65 eV [45]. Masato Wakamura et al. determined the effect of Ti substitution in HAp synthesized by the coprecipitation method on the bandgap which was compared with that of a typical anatase-TiO_2 photocatalytic powder. The experimentally obtained optical bandgap energies of TiHAp, HAp, and TiO_2 powder measured by diffuse reflectance spectroscopy were 3.65, >6, and 3.27 eV, respectively. The authors explained the increase of bandgap in TiHAp due to hybridizing of Ti 3d orbital with O 2p orbital and forming an internal state in the HAp bandgap, consequently, causing the absorption-edge lowering of TiHAp. Researching on the acetaldehyde gas decomposition of TiHAp by UV with VIS irradiation, the authors investigated the increase of activity comparing with when UV irradiation alone was used. Linh et al. [46] also indicated the shift of bandgap of TiO_2/HAp material synthesized by the hydrothermal process containing $TiO(OH)_2$ and HAp gels. The binding energy values of Ca 2p, P 2p, and O1s levels are related to hydroxyapatite phase, whereas those of Ti 2p levels corresponded with the characterization of titanium (IV) in TiO_2. The experimental bandgap of TiO_2/HAp material calculated by the DRS measurement was 3.6 eV, while those of pure HAp and TiO_2 were 5.3 and 3.2 eV, respectively. The shifts may relate oxidation states of titanium in the annealing process of $TiO(OH)_2$/HAp gel. The increase of bandgap may affect the absorption ability of radiation in the visible-light region; consequently, in some cases, it is a demoted impact which reduces the photocatalytic activity of TiO_2/HAp composites.

4. Conclusion

In the photocatalytic TiO_2/HAp composites, HAp plays the role of an adsorption of contaminations which is generally influenced by the physicochemical properties of HAp material. On the other hand, the promoted impacts of hydroxyapatite in the composites are experimentally collected in the specific cases including the absorption of radiation in the visible-light region, the small recombination rate of the electron-hole pair, and the high surface area of HAp component. However, the role of HAp in TiO_2/hydroxyapatite photocatalytic materials has been continuously researched and evaluated to obtain the best composite in the photocatalytic field, as well as in other applications.

Acknowledgements

This work is supported by Basic Science Research Programs (No. 2015R1D1A1A01056983) through the NRF, Korea, funded by the Ministry of Education and (No. 2018R1A2B6002194) by the Ministry of Science and ICT.

Author details

Linh Nguyen Thi Truc[1]*, Seungbum Hong[2] and Kwangsoo No[2]

*Address all correspondence to: linhntt@hcmup.edu.vn

1 Department of Chemistry, Ho Chi Minh City University of Education, Ho Chi Minh City, Vietnam

2 Department of Materials Science and Engineering, Korea Advanced Institute of Science and Technology, Daejeon, South Korea

References

[1] Fihri A, Len C, Varma RS, Solhy A. Hydroxyapatite: A review of syntheses, structure and applications in heterogeneous catalysis. Coordination Chemistry Reviews. 2017; **347**:48-76

[2] Hashimoto K. TiO_2 Photocatalysis: A historical overview and future prospects. Japanese Journal of Applied Physics. 2005;**44**(12):8269-8285

[3] Sandrini E, Giordano C, Busini V, Signorelli E, Cigada A. Apatite formation and cellular response of a novel bioactive titanium. Journal of Materials Science Materials in Medicine. 2007;**18**:1225-1237

[4] Yu P, Lu F, Zhu W, Wang D, Zhu X, Tan G, et al. Bio-inspired citrate functionalized apatite coating on rapid prototyped titanium scaffold. Applied Surface Science. 2014; **313**:947-953

[5] Aparicio C, Rodriguez D, Gil FJ. Variation of roughness and adhesion strength of deposited apatite layers on titanium dental implants. Materials Science and Engineering: C. 2011;**31**:320-324

[6] Chen D, Jordan EH, Gell M, Wei M. Apatite formation on alkaline-treated dense TiO_2 coatings deposited using the solution precursor plasma spray process. Acta Biomaterialia. 2008;**4**:553-559

[7] Kim H-W, Koh Y-H, Li L-H, Lee S, Kim H-E. Hydroxyapatite coating on titanium substrate with titania buffer layer processed by sol-gel method. Biomaterials. 2004;**25**:2533-2538

[8] Balamurugan A, Balossier G, Michel J, Ferreira JMF. Electrochemical and structural evaluation of functionally graded bioglass-apatite composites electrophoretically deposited onto Ti_6Al_4V alloy. Electrochimica Acta. 2009;**54**:1192-1198

[9] Chellappa M, Vijayalakshmi U. Electrophoretic deposition of silica and its composite coatings on Ti_6Al_4V, and its in vitro corrosion behaviour for biomedical applications. Materials Science & Engineering C-Materials for Biological Applications. 2017. Feb 1;**71**: 879-890

[10] Ievlev VM, Domashevskaya EP, Putlyaev VI, Tret'yakov YD, Barinov SM, Belonogov EK, et al. Structure, elemental composition, and mechanical properties of films prepared by radio-frequency magnetron sputtering of hydroxyapatite. Glass Physics and Chemistry. 2008;**34**(5):608-616

[11] Takemoto S, Yamamoto T, Tsuru K, Hayakawa S, Osaka A, Takashima S. Platelet adhesion on titanium oxide gels: Effect of surface oxidation. Biomaterials. 2004;**25**:3485-3492

[12] Tengvall P, Wälivaara B, Westerling J, Lundström I. Stable titanium superoxide radicals in aqueous Ti-peroxy gels and Ti-peroxide solutions. Journal of Colloid and Interface Science. 1991;**143**:589-592

[13] Unosson E et al. Stability and prospect of UV/H_2O_2 activated titania films for biomedical use. Applied Surface Science. 2013;**285**(Part B):317-323

[14] Reddy MP, Venugopal A, Subrahmanyam M. Hydroxyapatite photocatalytic degradation of calmagite (an azo dye) in aqueous suspension. Applied Catalysis B: Environmental. 2007;**69**(3-4):164-170

[15] Kawasaki T, Ikeda K, Takahashi S, Kuboki Y. Further study of hydroxyapatite high-performance liquid chromatography using both proteins and nucleic acids, and a new

technique to increase chromatographic efficiency. European Journal of Biochemistry. 1986;**155**:249

[16] Tanahashi M, Kamiya K, Suzuki T, Nasu H. Fibrous hydroxyapatite grown in the gel system: Effects of pH of the solution on the growth rate and morphology. Journal of Materials Science Materials in Medicine. 1992;**3**:48-53

[17] Yamasaki N, Kai T, Nishioka M, Yanagisawa K, Ioku K. Porous hydroxyapatite ceramics prepared by hydrothermal hotpressing. Journal of Materials Science. 1990;**9**:1150-1151

[18] Paschalis EP, Zhao Q, Tucker BE, Mukhopahayay S, Bearcroft JA, Beals NB, et al. Degradation potential of plasma-sprayed hydroxyapatite-coated titanium implants. Journal of Biomedical Materials Research. 1995;**29**:1499-1505

[19] Zhitomirsky I et al. Electrophoretic deposition of hydroxyapatite. Journal of Materials Science. 1997;**8**:213-219

[20] Xu S, Long J, et al. RF plasma sputtering deposition of hydroxyapatite bioceramics: Synthesis, performance, and biocompatibility. Plasma Process and Polymer. 2005;**2**: 373-390

[21] Nonami T, Hase H, Funakoshi K. Apatite-coated titanium dioxide photocatalyst for air purification. Catalysis Today. 2004;**96**:113-118

[22] Hirakura S, Kobayashi T, Ono S, Oaki Y, Imai H. Fibrous nanocrystals of hydroxyapatite loaded with TiO_2 nanoparticles for the capture and photocatalytic decomposition of specific proteins. Colloids and Surfaces B: Biointerfaces. 2010;**79**:131-135

[23] Katayama M, Tamazawa K, Kodaira A, Nonami T. Evaluation of protein adsorption capacity of TiO_2-supported spherical porous hydroxyapatite. Key Engineering Materials. ISSN: 1662-9795

[24] Ryu J, Kim K-Y, Hahn B-D, Choi J-J, Yoon W-H, Lee B-K, et al. Photocatalytic nanocomposit thin films of TiO_2-β-calcium phosphate by aerosol-deposition. Catalysis Communications. 2009;**10**:596-599

[25] Ma N, Fan X, Quan X, Zhang Y. Ag–TiO_2/HAp/Al_2O_3 bioceramic composite membrane: Fabrication, characterization and bactericidal activity. Journal of Membrane Science. 2009;**336**:109-117

[26] Xie J, Meng X, Zhou Z, Li P, Yao L, Bian L, et al. Preparation of titania/hydroxyapatite (TiO_2/HAp) composite photocatalyst with mosaic structure for degradation of pentachlorophenol. Materials Letters. 2013;**110**:57-60

[27] Komazaki Y, Shimizu H, Tanaka S. A new measurement method for nitrogen oxides in the air using an annular diffusion scrubber coated with titanium dioxide. Atmospheric Environment. 1999;**33**:4363-4371

[28] Ozeki K, Janurudin JM, Aoki H, Fukui Y. Photocatalytic hydroxyapatite/titanium dioxide multilayer thin film deposited onto glass using an rf magnetron sputtering technique. Applied Surface Science. 2007;**253**(7):3397-3401

[29] Liu Y, Yang Q, Wei JH, Xiong R, Pan CX, Shi J. Synthesis and photocatalytic activity of hydroxyapatite modified nitrogen-doped TiO_2. Materials Chemistry and Physics. 2011; **129**:654-659

[30] Kandoria K, Oketania M, Wakamura M. Decomposition of proteins by photocatalytic Ti(IV)-doped calcium hydroxyapatite particles. Colloids and Surfaces B: Biointerfaces. 2013;**102**:908-914

[31] Kandori K, Kuroda T, Wakamura M. Protein adsorption behaviors onto photocatalytic Ti(IV)-doped calcium hydroxyapatite particles. Colloids and Surfaces B: Biointerfaces. 2011;**87**: 472-479

[32] Kandori K, Oketani M, Sakita Y, Wakamura M. FTIR studies on photocatalytic activity of Ti(IV)-doped calcium hydroxyapatite particles. Journal of Molecular Catalysis A: Chemical. 2012;**360**:54-60

[33] Hu M, Yao Z, Liu X, Ma L, He Z, Wang X. Enhancement mechanism of hydroxyapatite for photocatalytic degradation of gaseous formaldehyde over TiO_2/hydroxyapatite. Journal of the Taiwan Institute of Chemical Engineers. 2018;**85**:91-97

[34] Mohamed RM, Baeissa ES. Preparation and characterisation of Pd–TiO_2–hydroxyapatite nanoparticles for the photocatalytic degradation of cyanide under visible light. Applied Catalysis A: General. 2013;**464-465**:218-224

[35] Zhang X, Yates MZ. Enhanced photocatalytic activity of TiO_2 nanoparticles supported on electrically polarized hydroxyapatite. ACS Applied Materials & Interfaces. 2018; **10**(20):17232-17239

[36] Kobayashi A, Jiang W. Properties of titania/hydroxyapatite nanostructured coating produced by gas tunnel type plasma spraying. Vacuum. 2009;**83**:86-91

[37] Liu Y, Liu CY, Wei JH. Enhanced adsorption and visible-light-induced photocatalytic activity of hydroxyapatite modified Ag-TiO_2 powders. Applied Surface Science. 2010; **256**(21):6390-6394

[38] Aramendia MA, Hidalgo-Carrillo J, Sebti J. A study on the potential application of natural phosphate in photocatalytic processes. Journal of Colloid and Interface Science. 2010;**344**(2):475-481

[39] Nishikawa M, Yang W, Nosaka Y. Grafting effects of Cu^{2+} on the photocatalytic activity of titanium-substituted hydroxyapatite. Journal of Molecular Catalysis A: Chemical. 2013;**378**:314-318

[40] Li Q, Feng X, Zhang X, Song H, Zhang J, Shang J, et al. Photocatalytic degradation of bisphenolusing Ti-substituted hydroxyapatite. Chinese Journal of Catalysis. 2014; **35**:90-98

[41] Rosenman G, Aronov D, Oster L, et al. Photoluminescence and surface photovoltage spectroscopy studies of hydroxyapatite nano-bio-ceramics. Journal of Luminescence. 2007;**122-123**(1-2):936-938

[42] Rulis P, Ouyang L, Ching WY. Electronic structure and bonding in calcium apatite crystals: Hydroxyapatite, fluorapatite, chlorapatite, and bromapatite. Physical Review B—Condensed Matter and Materials Physics. 2004;**70**(15):155104

[43] Matsunaga K, Kuwabara A. First-principles study of vacancy formation in hydroxyapatite. Physical Review B. 2007;**75**. Article ID 014102:1-9

[44] Mitsionis A, Vaimakis T, Trapalis C, Todorova N, Bahnemann D, Dillert R. Hydroxyapatite/titanium dioxide nanocomposites for controlled photocatalytic NO oxidation. Applied Catalysis B: Environmental. 2011;**106**(3-4):398-404

[45] Tsukada M, Wakamura M, Yoshida N, Watanabe T. Band gap and photocatalytic properties of Ti-substituted hydroxyapatite: Comparison with anatase-TiO_2. Journal of Molecular Catalysis A: Chemical. 2011;**338**:18-23

[46] Linh NTT, Tuan PD, Dzung NV. The shifts of band gap and binding energies of titania/hydroxyapatite material. Journal of Composites. 2014. Article ID 283034;**2014**:5

Section 2

Photocatalyst for Photoreduction

Chapter 8

Photoreduction Processes over TiO_2 Photocatalyst

Endang Tri Wahyuni and Nurul Hidayat Aprilita

Additional information is available at the end of the chapter

http://dx.doi.org/10.5772/intechopen.80914

Abstract

This chapter presents the study of TiO_2 photocatalyst for the photoreduction of several reducible chemicals. The photocatalytic reduction of several toxic metal ions, including Ag (I), Cu(II), Cr(VI), Hg(II), and U(VI) in the presence of TiO_2, in order to decrease their toxicity, is described. Photodeposition of the noble metals, such as Ag(I), Au(III), Pt(IV), and Pd(II) for doping purposes by photocatalytic reduction over TiO_2, is also addressed. Conversion of the greenhouse gas of CO_2 into useful hydrocarbons and methanol by photocatalytic reduction using TiO_2 photocatalyst is highlighted. Several operating parameters in photoreduction processes that are photocatalyst dose, time of the irradiation, pH of the solution, and the initial concentration of the substrates (the reducible chemicals) are also reviewed.

Keywords: photoreduction, TiO_2 photocatalyst, detoxification, doping, CO_2 conversion

1. Introduction

Semiconductor TiO_2 is regarded as one of the most promising photocatalysts, because it has low cost and high activity, good physical and chemical stability, and nontoxic property [1]. The structure of a semiconductor is characterized by two bands, called as valence band and conduction band, that are separated by a gap named as bandgap energy (Eg). The first band is filled by electrons, while the second is empty or no electron occupying it. The bandgap of TiO_2 with anatase typed is 3.2 and 3.0 eV is for the rutile type [2]. Both anatase and rutile are tetragonal in structure, but the anatase has octahedrons that share four edges forming the fourfold axis.

The photocatalyst of TiO_2 works by absorption of UV to near visible region, with the energy same as or higher than its bandgap (light with a wavelength of 365 nm is required by rutile and that of 411 nm is for anatase) that generates electron and hole pair. The pair generation is resulted by excitation of some electrons from valence band (e_{vb}) to the conduction one (e_{cb}) with the formation of positive holes (h_{vb}+) [3, 4]. The reactions of electron–hole generation from TiO_2 are presented as Eqs. (1):

$$TiO_2 + hv \rightarrow TiO_2\left(e^- + h^+\right) \quad (1)$$

$$h^+ + H_2O \rightarrow \cdot OH + H^+ \quad (2)$$

$$h^+ + TiOH \rightarrow Ti\,OH \quad (3)$$

$$e^- + O_2 \rightarrow O_2{}^- \quad (4)$$

$$TiO_2 + \left(e^- + h^+\right) \rightarrow TiO_2 + heat \quad (5)$$

Notes: *hv* represents photon energy of UV light ($E = hv$), $O_2{}^-$ is called as superoxide, and ·OH indicates hydroxide radical.

The electrons, in water media and in the presence of dissolve oxygen, can react with the oxygen to form super oxide, as presented by reaction (4) [4]. This is a reduction path. The use of the electrons from TiO_2 for some reducible metal ions such as Ag(I) [5–13], Au(III) [13–17], Cr(VI) [18–23], Cu (II) [24–26], Hg(II) [27, 28], and U(VI) [29, 30] has also been developed.

Meanwhile the hole generates the free OH radical after contact with water and TiO_2 surface, as illustrated by Eq. (2) and (3). The OH radical acts as a strong oxidizing agent with high oxidation potential (E^0 = 2.8 V) that can degrade organic compounds into smaller CO_2 and H_2O molecules [3, 4]. This is called as oxidation path. Due to strong ability in organic degradation, OH radicals from TiO_2 have been intensively applied for cyanide oxidation [31], treating hazardous phenol [32, 33], di-nitrophenol [34, 35], various organic dyes [36–39], and surfactant of detergent [40].

The applications of the photodegradation process catalyzed by TiO_2 have been frequently published through journals and books. Meanwhile, less intensive photoreduction methods are explored. In fact, the photoreduction process over TiO_2 has some advantages compared to the other reduction reactions, with respect to simplicity, cost-effectiveness, efficiency, less chemical usage, and green chemistry principles.

It is interesting therefore to address the applications of the reduction over TiO_2 photocatalyst for various purposes. The applications of the photocatalytic reduction in the presence of TiO_2 for detoxifying the toxic metal ions, doping by transition and noble metals, and converting greenhouse gas CO_2 into more valuable chemicals are described in more detail.

2. Photocatalytic reduction over TiO_2 to detoxify metal ions

The reducible hazardous heavy metal ions that are widely found in the wastewater are silver Ag(I), copper Cu(II), hexavalent chromium Cr(VI), mercury Hg(II), and uranium U(VI). Silver pollutant can be found with high concentration in the wastewater of radiophotography activity that is usually disposed from hospitals [41, 42], silver electroplating, and electronic fabrication [43].

Wastewaters containing copper in high concentration are disposed from electroplating and electrical stuffs [43, 44]. Hexavalent chromium is mostly emitted by metal plating activity and paint industry [45, 46]. Mercury is usually contained in wastewater originated from gold recovery and incinerator in hospitals [46, 47]. The environmental contamination by uranium (VI) ion may originate from uranium purification or extraction from its respected mineral. A leak of nuclear reactor releasing uranium into the water may also contribute to the uranium contamination [48, 49].

Contamination of the hazardous metals can create environmental and human health problems. Silver contamination can induce argyria syndrome [50]. For human, the excessive copper intake can disturb the gastrointestinal (GI) system [44]. Hexavalent chromium is carcinogenic [45, 51], and mercury Hg(II) can cause neurological dysfunction [47]. The uranium ion, as a radioactive element, must be very dangerous, both for environment and human [44]. Therefore, a method that can detoxify the hazardous heavy metals is urgently.

The products of the reduction of Ag(I), copper Cu(II), hexavalent chromium Cr(VI), mercury Hg(II), and uranium (VI) are Ag(0), Cu(0), Cr(III), Hg(0), and U(IV), respectively, that are less toxic [44]. Hence, these facts motivate many researchers for detoxification of the hazardous heavy metal ions by reduction method, especially by photoreduction catalyzed by TiO_2.

Photoreduction of Ag(I) ions in the aqueous solution by electron provided by TiO_2 takes place effectively, due to its high standard reduction potential (E^0) = 0.79 V [52]. The reaction is presented by Eq. (6) that produces undissolved solid silver that is less toxic and easier to be handled. It is clear that by photoreduction process, the silver contaminant is detoxified:

$$Ag^+ + e^- \rightarrow Ag^0 \qquad E^0 = 0.79\ V \tag{6}$$

Hexavalent chromium in the aqueous solution existed as chromate ($CrO_4^=$) and bichromat ($Cr_2O_7^=$) ions, depending on the pH that has high standard reduction potential (E^0) = 1.33 V in acid condition [52]. This allows them to be easily reduced into Cr(III) as Cr^{3+} ions as presented by reaction (7). The chromate (Cr(VI)) is highly toxic, while Cr^{3+} is less toxic or is even needed by feeding women [45]. Hence, detoxification of the toxic Cr(VI) can be carried out by reduction method. Photoreduction of Cr(VI) over TiO_2 photocatalyst has been frequently reported [18–23] with satisfaction result. Further, the Cr^{3+} ions can precipitate into undissolved $Cr(OH)_3$ in basic condition that is remediable by solidification/stabilization technique:

$$Cr_2O_7^= + 14H^+ + e^- \rightarrow 2Cr^{3+} + 7H_2O \qquad E^0 = 1.52\ V \tag{7}$$

Copper ions in the solution formed as Cu^{2+} can be reduced into dissolved Cu (I) and/or undissolved Cu(0), with respective E^0 = 0.153 V and 0.34 V [52]. Photoreduction method in the presence of TiO_2 has been subjected to detoxify the toxic Cu^{2+} that can prominently form the undissolved toxic Cu(0), with very small Cu(I) ion [24–26]. The reactions of the Cu(II) reduction are shown by Eqs. (8) and (9). The photoreduction of Cu(II) is found to be less effective compared to Ag(I) photoreduction that may be caused by the low standard reduction potential, E^0 = 0.34 V [52]. To improve the effectiveness, a reducing agent, such as oxalic acid, can be added in the Cu(II) photoreduction [26]:

$$Cu^{2+} + e \rightarrow Cu^{+} \qquad E^0 = 0.153\ V \tag{8}$$

$$Cu^{2+} + 2\,e \rightarrow Cu^{0} \qquad E^0 = 0.34\ V \tag{9}$$

Photoreduction over TiO_2 has been also used to detoxify mercury (Hg^{2+}) ion in the aqueous solution, by converting it to be undissolved Hg^0 [27, 28]. Based on the standard reduction potential as seen in the reaction (10) [52], the reduction should proceed effectively. To handle the elemental or solid mercury may be easier than that of the dissolved ions. As presented by previous authors [47], the order of the toxicity level of mercury forms, from the most toxic, is methyl mercury (CH_3Hg), $Hg^{(0)}$ vapor, Hg^{2+} dissolved ion, and $Hg^{(0)}$ element:

$$Hg^{2+} + 2e^{-} \rightarrow Hg^{0} \qquad E^0 = 0.85\ V \tag{10}$$

The photoreduction catalyzed by TiO_2 suspension has also been studied for removal of the radioactive uranium (VI) [48, 49] that exists as ${UO_2}^{2+}$ anionic in the solution. The anion is the most stable form and so the one that is found in the solution. The photoreduction of the anionic has produced the less radioactive uranium (V), Eq. (11) [52]:

$${UO_2}^{2+} + e^{-} \rightarrow {UO_2}^{+} \qquad E^0 = 0.163\ V \tag{11}$$

Detoxification of the hazardous (toxic and radioactive) heavy metals by photoreduction pathway offers a simple, practical, economic, and green procedure that meets with the future requirement method in solving the environmental pollution problem.

3. Photoreduction method over TiO_2 for metal recovery and doping purposes

Photoreduction of some reducible metal ions results in the solid phase deposited on the TiO_2 structure. This deposition phenomenon has inspired some researchers to use the photoreduction for metal recovery and doping purposes.

Recovery is objected to get pure valuable metals such as silver (Ag) and gold (Au) from the regarding solution, by applying the photoreduction method. The photoreduction of Ag(I) follows Eq. (6). To take the pure Ag particles from Ag deposited on TiO_2, one can use ultrasonic shaker [12].

Photodeposition for gold recovery can also be obtained through reduction under UV light in the presence of TiO_2 photocatalyst [17]. In this step, the gold is dissolved in the aqueous solution formed as Au^{3+} ions and then to be reduced as shown by reaction (12):

$$Au^{3+} + 3e^{-} \rightarrow Au^{0} \qquad E^0 = 1.32\ V \tag{12}$$

Doping, whether with transition and noble metals on TiO_2 recently, has attracted much attention, since it can improve the performance of TiO_2 as a photocatalyst under UV light, as well as shift the absorption of TiO_2 to visible light region [5–11]. The latter is supposed to give some advantages, as the photocatalytic process under metal-doped TiO_2 can take place under sun light that must be low-cost and safer and so greener than that of by UV light irradiation.

The transition metals that have been examined as dopants on TiO_2 are Cu(II) [53, 54], Fe [55, 56], Co [56, 57], Ni [57], Mn [56], and Cr [58]. Moreover, Ag(I) [5–13], Au(III) [13, 14], Pd(II) [14], and Pt(IV) [14] are the noble metals that are frequently doped into TiO_2 structure. All the metals doped on TiO_2 are reported to improve the photocatalytic activity of TiO_2 under UV irradiation as well as to shift the visible absorption with various effects, from very low, shown by Cr(III) up to the very significant effect, observed on Ag(I).

A doping process basically involves the conversion of the metal ions in the solution to be deposited solid metal on TiO_2 powder that is frequently carried out by sol–gel method [56]. However, hydrothermal [57] and chemical vapor [58] methods are also introduced.

From the above methods, the regard salt solution is usually used as the dopant source, and high temperature is required that makes the method costly due to high energy consumption. In addition, large metal particle is usually resulted from the process that retards the metal insertion into gap of valence and conduction gaps. As a consequence, the small absorption shift is resulted that yields less significant improvement of the photocatalyst activity under UV light or the slight visible light responsiveness.

In addition to the four doping methods, photoreduction has also been examined. The photoreduction method becomes a great of interest, because the process takes place at room temperature, no need of chemicals, except UV light, and has resulted small cluster of metal dopant particles. The small particles are well inserted into the gap between valence and conduction band of TiO_2. Such insertion has considerably shortened the gap that enhances the activity under UV light and pronounces shift of the light absorption into wide visible region. However, the photoreduction method is limited only for dopants that are reducible metal ions, including Cu(II) representing transition metal and Ag(I), Au(III), Pd(II), and Pt(IV) for noble metal ions.

In general, the doping process by photoreduction method is carried out by UV light irradiation toward the regard metal salt solutions in a certain period of time. Then M/TiO_2 (M = metal dopant) resulted is dried at about 110°C to remove the water.

Photoreduction of Ag (I) in the solution over TiO_2 for doping purpose principally follows the same procedure as in the detoxification, as described earlier. The starting salt for Ag doping usually used is $AgNO_3$ [5–13].

As its high standard reduction potential (E^0), the photoreduction of Ag^+ takes place efficiently, and the small Ag particle resulted can enter into the gap between the conduction and the valence. The present of the small particle dopant in the gap shortens the bandgap. This allows the metal-doped TiO_2 to be active under visible light and to work better with UV irradiation, whether for degradation of the organic pollutants or for bacterial combating.

In the doping Au on TiO_2 through photoreduction method, the salt frequently introduced as gold source is $KAuCl_4$ that dissolves to form $AuCl_4^-$ [13–16]. The doping follows reaction (13). The other gold ions may form as $AuCl_2^-$, Au^+, and Au^{3+} that are also reducible by the following reactions (14)–(16) with their own standard reduction potentials [52]:

$$AuCl_4^- + 3e^- \rightarrow Au^0 + 2Cl_2 \qquad E^0 = 0.93\ V \tag{13}$$

$$Au^+ + e^- \rightarrow Au^0 \qquad E^0 = 1.83\ V \tag{14}$$

$$Au^{3+} + 3e^- \rightarrow Au^0 \qquad E^0 = 1.32\ V \tag{15}$$

$$AuCl_2^- + e^- \rightarrow Au^0 + Cl_2 \qquad E^0 = 1.15\ V \tag{16}$$

The gold atom resulted from the photoreduction is doped on TiO_2 structure or TiO_2/Au through insertion or impregnation. The doped photocatalyst has been tested for phenol degradation under UV light and showed more satisfaction result than undoped TiO_2 [13].

Platinum (Pt) doped TiO_2 or TiO_2/Pt can be resulted by irradiating H_2PtCl_6 salt in the aqueous solution in the presence of TiO_2 suspension [14]. The platinum salt is dissolved to form an ion of $PtCl_4^{4+}$ and/or Pt^{2+} ions, and the reactions of the photoreduction are represented by Eqs. (17) and (18) [52]:

$$PtCl_4^{2-} + 2e^- \rightarrow Pt^0 + 2Cl_2 \qquad E^0 = 0.77\ V \tag{17}$$

$$Pt^2 + +2e \rightarrow Pt^0 \qquad E^0 = 1.20\ V \tag{18}$$

Photoreduction of Pd(II) over TiO_2 for the doping purpose is performed by the irradiation of $PdCl_2$ in the aqueous solution with UV light [14]. The reduction of Pd(II) ion is written as reaction (19) that yields small Pd particles on TiO_2 structure:

$$Pd^2 + +2e^- \rightarrow Pd^0 \qquad E^0 = 0.915\ V \tag{19}$$

With the same conditions of the photoreduction, it is found that the order of the photodeposition efficiency from the highest is shown by Pt (100%) that is followed by Au (80%) and then by Pd (50%). This sequence photoreduction result is consistent with their standard reduction potentials that are 1.2, 0.93, and 0.915 V, as well as their empirical radii that are 135, 135, and 140, respectively. The higher standard reduction potentials promote more reduction, and smaller size facilitates the effective insertion. They have also been examined for phenol degradation and displayed the effective results [14].

4. Photoreduction method over TiO_2 of the greenhouse gas CO_2 into valuable substances

The level of CO_2 that is a primary greenhouse gas in the atmosphere is continuously rising that creates serious global warming. Conversion of CO_2 into more valuable compounds is believed to be the best way in preventing the excess CO_2 disposal in the air. CO_2 is a thermodynamically stable and chemically inert compound, and it is difficult to oxidize or to reduce it to other useful compounds under normal operating conditions.

http://dx.doi.org/10.5772/intechopen.80914

Therefore, converting CO_2 into valuable products is possible when catalytic, electrocatalytic, plasmatic, enzymatic, and photocatalytic reduction processes [59] are employed. Among them, photocatalytic reduction seems to be the most intensively developed method.

The photoreduction of CO_2 with water vapor catalyzed by titania-based photocatalysts results in methane (CH_4), methanol (CH_3OH), carbon monoxide (CO), formic acid (HCOOH), and formaldehyde (HCHO) and follows the simplified reactions (20)–(23) [59].

$$2CO_2 + 4H_2O \rightarrow 2CH_3OH + 3O_2 \quad (20)$$

$$3CO_2 + 2H_2O \rightarrow CH_4 + 2CO + 3O_2 \quad (21)$$

$$CO_2 + 2H_2O \rightarrow 2HCOOH + O_2 \quad (22)$$

$$CO_2 + 2H_2O \rightarrow HCOH + O_2 \quad (23)$$

The conversion reaction pathways are not specific and mainly depend on the reaction conditions. This is therefore a complex mechanism that proceeds through branching pathways and produces different products simultaneously [59].

Reduction of CO_2 in the presence of NaOH solution photocatalyzed by TiO_2 supported on a polymer has been reported to produce methanol and methane, accompanied with formic acid and formaldehyde. The CO_2 is being soluble in NaOH solution and forms carbonate and bicarbonate ions based on the pH measurement. The reductions of carbonate acid and carbonate ions with their standard reduction potential are shown as Eqs. (24) and (25) [60]:

$$H_2CO_3 + 6H^+ + 6e^- \rightarrow CH_3OH + 2H_2O \qquad E^0 = 0.044\ V \quad (24)$$

$$CO_3^{2-} + 8H^+ + 6e^- \rightarrow CH_3OH + 2H_2O \qquad E^0 = 0.209\ V \quad (25)$$

From the equations, based on the standard reduction potential reduction, the photoreduction of carbonate ions (mostly existing in higher pH) to form methanol takes place faster or is more effective than the photoreduction of the carbonate acid.

Various mechanistic reaction schemes have been proposed for CO_2 reduction with H_2O using TiO_2 photocatalysts. The following are the reaction mechanisms proposed for methane formation [61]:

$$H_2O + 2h^+ \rightarrow \cdot OH + H\cdot \quad (26)$$

$$CO_2 + e^- \rightarrow CO_2\cdot \quad (27)$$

$$2CO_2\cdot + 2e \rightarrow 2CO\cdot + O_2 \quad (28)$$

$$2CO\cdot + 4H\cdot \rightarrow 2\cdot CH_2 + O_2 \quad (29)$$

$$2\cdot CH_2 + 2H\cdot \rightarrow 2CH_4 \quad (30)$$

The above mechanism steps start with the reaction between water molecule with the hole or positive radical from TiO_2 (Eq. (1)) to form radicals of OH and H (Eq. (26)). At the same time, carbon dioxide is reduced by the electron from Eq. (1), to form negative radical of carbon dioxide (Eq. (27)). The further reduction of CO_2 radical is preceded into carbon monoxide radical and oxygen (Eq. (28)). The CO radical then reacts with hydrogen radical from Eq. (26) to produce radical of methylene (Eq. (29)). Finally the methylene radical reacts with hydrogen radical to yield methane (Eq. (30)).

Different mechanisms for hydrogen, carbon monoxide, methane, and methanol productions are also proposed as seen in Eqs. (31)–(33) [62]. In this mechanism, only electron plays a role in the photoreduction, while no involvement of the hole is illustrated. Firstly, carbon dioxide gas reacts with the electron released from TiO_2 (Eq. (1)), to form anion of carbon dioxide (Eq. (31)). At same time, the hydrogen ion from water is also reduced by the electron to form hydrogen gas (Eq. (32)). Secondly, the anion of carbon dioxide reacts with hydrogen ion in the presence of electron to result carbon monoxide and water (Eq. (33)). In addition, the reaction of carbon dioxide anion with hydrogen ion and electron also occurred to produce methane and oxygen (Eq. (34)). Simultaneously, the carbon dioxide anion also reacts with the hydrogen ion and electron to yield methanol and water (Eq. (35)):

$$CO_2 + e^- \rightarrow CO_2^- \tag{31}$$

$$2H^+ + e^- \rightarrow H_2 \tag{32}$$

$$CO_2^- + 2H^+ + e^- \rightarrow CO + H_2O \tag{33}$$

$$CO_2^- + 2H^+ + e^- \rightarrow CH_4 + O_2 \tag{34}$$

$$CO_2^- + 2H^+ + e^- \rightarrow CH_3OH + H_2O \tag{35}$$

The methane and hydrogen gas resulted from photoreduction of CO_2 with H_2O under UV light and over TiO_2 is also formulated by mechanisms as follows [63]. The reactions between water molecule with a hole or positive radical photogenerated by TiO_2 under UV light irradiation (Eq. (36)) yield OH radicals and H ion (Eq. (37)). Then the OH radical further reacts with water and the positive radical, to release oxygen gas and hydrogen ion (Eq. (38)). The overall of the reactions (37) and (38) is represented by Eq. (39):

$$TiO_2 + \mathit{UV\ light} \rightarrow TiO_2 + e^- + h^+ \tag{36}$$

$$H_2O + h^+ \rightarrow \cdot OH + H^+ \tag{37}$$

$$\cdot OH + H_2O + 3h^+ \rightarrow O_2 + 3H^+ \tag{38}$$

$$2H_2O + 4h^+ \rightarrow O_2 + 4H^+ \tag{39}$$

On the other side, CO_2 meets with the electron to form a corresponding radical (Eq. (40)). Then the radical is dissociated into carbon monoxide radical along with oxygen (Eq. (41)). The carbon monoxide radical is further split into carbon radical and oxygen (Eq. (42)). The overall reaction is written as Eq. (43):

$$CO_2 + e \rightarrow \cdot CO_2 \quad (40)$$

$$\cdot CO_2 \rightarrow \cdot CO + 1/2O_2 \quad (41)$$

$$\cdot CO \rightarrow \cdot C + 1/2O_2 \quad (42)$$

$$CO_2 \rightarrow \cdot C + O_2 \quad (43)$$

The hydrogen ion obtained from the reaction (39) then reacts with the carbon radical from reaction (43) to yield methane and hydrogen gas, as represented by Eqs. (44) and (45), respectively:

$$4H^+ + \cdot C + 4e \rightarrow CH_4 \quad (44)$$

$$2H^+ + 2e \rightarrow H_2 \quad (45)$$

The following is the other mechanism in forming methanol, methane, and ethylene that is proposed [64–66]. After being released from TiO_2 (Eq. (46)), the hole reacts with water molecule to form radicals of hydroxyl and hydrogen (Eq. (47)), while the electron reduces the carbon dioxide to yield its anionic form (Eq. (48)). Then the anion of carbon dioxide reacts with the hydrogen radical from Eq. (47), to result an intermediate radical and hydroxide anion (Eq. (49)):

$$TiO_2 + UV\ light \rightarrow TiO_2 + e^- + h^+ \quad (46)$$

$$2H_2O + 2h^+ \rightarrow \cdot OH + H\cdot \quad (47)$$

$$CO_2 + e^- \rightarrow {CO_2}^- \quad (48)$$

$${CO_2}^- + 2H\cdot \rightarrow OC\cdot H + OH^- \quad (49)$$

$$OC\cdot H + OC\cdot H \rightarrow HOCCOH \quad (50)$$

$$HOCCOH + 4H\cdot \rightarrow CH_3COH \quad (51)$$

$$CH_3COH + H\cdot \rightarrow \cdot CH_3 + CO \quad (52)$$

$$\cdot CH_3 + H\cdot \rightarrow CH_4 \quad (53)$$

$$\cdot CH_3 + \cdot CH_3 \rightarrow C_2H_6 \quad (54)$$

The product of the radical (Eq. (50)) reacts with hydrogen radical to produce acetic acid (Eq. (51)). The acid can also further react with hydrogen radical to form methyl radical (Eq. (52)). Next, the methyl radical reacts with hydrogen to produce methane. When two methyl radicals react to each other, ethane is produced (Eq. (54)).

5. Variables influencing the photoreduction process

In the study of the photocatalytic reduction method, it is found that some variables including photocatalyst dose, reaction time, UV lamp types, pH, and concentration of the reducible ions considerably play an important role in the photoreduction results.

5.1. The influence of the photocatalyst dose

In general, the increase of the photocatalyst dose promotes higher photoreduction efficiency, and then it declines when the photocatalyst dose is further increased. Such trend can be explained in terms of the number of active sites available for photocatalytic reactions. The larger number of the active sites is available in as the dose of the photocatalyst increases that would generate more numbers of electrons and so higher photoreduction effectiveness. However, the large amount of catalyst may result in agglomeration form with larger particle size [40] that may provide smaller surface area. The agglomeration can also induce light scattering that reduces the light contacting with TiO_2 surface.

The smaller surface area can reduce the number of active sites and so decreases the electron number provided. Consequently lower photoreduction is obtained. Another reason of the decrease in the photoreduction can be attributed to the increase in the turbidity of suspension due to the large amount of photocatalyst. This leads to the inhibition of photon absorption by the photocatalyst. As an effect, the lesser photoinduced electrons can be provided, causing the photoreduction decreased [23]. From several studies, the optimum photocatalyst dose was reported to be 2 g/L (Ag(I)) [8], 1.6 g/L (Cr(VI)) [20], 0.1 g/L (Cu(II)) [25], and 2 g/L (Hg(I)) [27].

5.2. The influence of the reaction time

The reaction time determines how long is the contact between light with photocatalyst and that of electrons with the reducible metal ions. The general trend observed is that the photoreduction efficiency improves as the ratio of time is further extended. Longer than the optimum time, the photoreduction is usually independent on the reaction time. In the beginning and further extension time, the contact between light and TiO_2 becomes more effective, resulting in more number of electrons. Then, the extension time allows more effective contact between the electrons available with the reducible metal ions. This can enhance significantly the photoreduction. At one time, the photoreduction reaches maximum level, showing the optimum reaction time. For longer time than the optimum one, a very large amount of the products has been resulted that may prevent the contact among the reacting agents. Consequently, TiO_2 is hindered to release more electrons that give no increase in the photoreduction. The other reason is the reducible ions in the solution have been completely reduced that no more ions are left in the solution. The optimum reaction time is detected to be varied: 50–60 min and 5 h for Ag(I) [5, 10], 4 h for Cr(VI) [23], 3 h for Cu(II) [25], and 50–150 min for Hg (II) [27].

http://dx.doi.org/10.5772/intechopen.80914

5.3. The influence of pH

The other important variable is pH, since pH determines the species of both TiO_2 surface and the reducible metal ions that further affect in the photoreduction efficiency [12]. The general trend of the metal ion photoreduction efficiency with the alteration pH is that at the low pH, the photoreduction is usually low, and then increasing pH in the acid range gives a rise in the photoreduction results, but the further increasing pH leads to the photoreduction decreased. In the aqueous media, the low pH provides more amount of hydrogen ion H^+ that can interact with TiOH (TiO_2 hydrated) surface to form $Ti \cdot OH_2^+$. Such protonated titanol $Ti \cdot OH_2^+$ is more difficult to release electrons [2–4], although the metal ions mostly existed as the species that are easier to be reduced in large amount; the low photoreduction efficiency is usually observed. It is clear that the number of the electrons plays a prominent role on the photoreduction process.

Increasing pH in the acid range, the H^+ concentration declines that make TiOH available increased. This can raise the release of the electrons, and the metal ions are found as reducible species, so the photodegradation can be considerably enhanced. Finally, the number of OH may be in excess at higher pH that can create negative surface of TiO_2 to form TiO^-. This can retard the electron released. On the other side, generally the metal ions are precipitated with the hydroxide anions to form $M(OH)_n$ (M = metal, n = valence) that is impossible to be reduced. These two situations stimulate the negative effect on the photoreduction. As the other variables, there are also optimum pH values that are varied depending on the respective metal ions. Optimum pH value for Ag is found at 6–7 [8], at pH = 2 for Cr(VI) [23], and for Cu(II) is 2–3.5 [26], and pH = 4–4.1 is reported for Hg(II) [27]. The photoreduction of CO_2 with maximum results is observed at pH 4 [40].

5.4. The influence of the initial concentration

The initial concentration of the reducible metal ions is also important to be paid of attention in the photoreduction. In general, the relationship of the initial concentration is as follows: in the increase of the initial concentration, the metal ions in the photoreduction are enhanced. When the amount of the reduced ions is compared/relative to the initial concentration, it is usually observed that the photoreduction percentage decreases as the initial concentration increases. With the low initial concentration, the small number of the reducible ions is present with abundant electrons that can be reduced completely or 100%. By increasing the initial concentration of the metal ion, the number of the metal ions to be reduced is also increased, but the percentage is usually decreased. The optimum initial concentrations are reported to be 4×10^{-3} mole/L for Ag (I) [8], 3×10^{-3} mole/L for Cu(II) [24], 10 mg/L for Cr(VI) [23], 100 mg/L for Hg(II) [28], and the CO_2/H_2O ratio is 1/2 [66].

6. Conclusions

Photocatalytic reduction over TiO_2 has shown as an attractive method that can be applied for solving the environmental problems due to the toxic metal ion contamination, providing new and renewable energy by conversion of CO_2 into syngas, improvement TiO_2 photocatalyst activity under UV and solar light by doping process, and recovery of valuable economic metals based on the metal deposited.

Author details

Endang Tri Wahyuni* and Nurul Hidayat Aprilita

*Address all correspondence to: endang_triw@ugm.ac.id

Chemistry Department, Faculty of Mathematic and Natural Sciences, Gadjah Mada University, Indonesia

References

[1] Fujishima A, Rao TN, Tryk DA. Titanium dioxide photocatalysis. Journal of Photochemistry and Photobiology C. 2000;**1**:1-21

[2] Hoffmann MR, Martin ST, Choi W, And Bahnemann DW. Environmental Applications of Semiconductor Photocatalysis. Chemical Reviews. 1995;**95**:69-96

[3] Linsebigler AL, Lu G, Yates JT. Photocatalysis on TiO_2 surfaces: Principles, mechanisms, and selected results. Chemical Reviews. 1995;**95**:735-758

[4] Nakata K, Fujishima A. TiO_2 photocatalysis: Design and applications. Journal of Photochemistry and Photobiology C: Photochemistry Reviews. 2012;**13**:169-189

[5] Seery MK, George R, Floris P, Pillai C. Silver doped titanium dioxide nanomaterials for enhanced visible light photocatalysis. Journal of Photochemistry and Photobiology A: Chemistry. 2007;**189**:258-263

[6] Sung-Suh HM, Choi JR, Hah HJ, Koo SM, Bae YC. Comparison of Ag deposition effects on the photocatalytic activity of nanoparticulate TiO_2 under visible and UV light irradiation. Journal of Photochemistry and Photobiology A: Chemistry. 2004;**163**:37-44

[7] Guin D, Manorama SV, Latha JNL, Singh S. Photoreduction of silver on bare and colloidal TiO_2 nanoparticles/nanotubes: Synthesis, characterization, and tested for antibacterial outcome. Journal of Physical Chemistry C. 2017;**111**:13393-13397

[8] Anandan P, Kumar S, Pugazhenthiran N, Madhavan J, Maruthamuthu P. Effect of loaded silver nanoparticles on TiO_2 for photocatalytic degradation of Acid Red 88. Solar Energy Materials & Solar Cells. 2008;**92**:929-937

[9] Sobana N, Muruganadhan M, Swaminathan M. Nano-Ag particles doped TiO_2 for efficient photodegradation of Direct azo dyes. International Journal of Molecular Sciences. 2006;**13**:13275-13293

[10] Sobana N, Selvam K, Swaminathan M. Optimization of photocatalytic degradation conditions of Direct Red 23 using nano-Ag doped TiO_2. Separation and Purification Technology. 2008;**62**:648-653

[11] Kumar R, Rashid J, Barakat MA. Zero valent Ag deposited TiO_2 for the efficient photocatalysis of methylene blue under UV-C light irradiation. Colloids and Interface Science Communications. 2015;**2**:1-4

[12] Wahyuni ET, Aprilita NH, Mudasir M. Influence of Cu(II) on Ag(I) recovery by photocatalytic reduction method with TiO_2 suspension. Journal of Chemistry and Chemical Engineering. 2013;**4**(9):50-53

[13] Chan SC, Barteau MA. Preparation of highly uniform Ag/TiO_2 and Au/TiO_2 supported nanoparticle catalysts by photodeposition. Langmuir. 2005;**21**:5588-5595

[14] Maicu M, Hidalgo MC, Colón G, Navío JA. Comparative study of the photodeposition of Pt, Au and Pd on pre-sulphated TiO_2 for the photocatalytic decomposition of phenol. Journal of Photochemistry and Photobiology A: Chemistry. 2011;**217**:275-283

[15] Hidalgo MC, Murcia JJ, Navío JA, Colón G. Photodeposition of gold on titanium dioxide for photocatalytic phenol oxidation. Applied Catalysis A: General. 2011;**397**:112-120

[16] Zhang D. Photocatalytic applications of Au deposited on TiO_2 nanocomposite catalyst in dye degradation via photoreduction. Russian Journal of Physical Chemistry A. 2012;**86**(3):498-503

[17] Wahyuni ET, Kuncaka A, Sutarno S. Application of photocatalytic reduction method with TiO_2 for gold recovery. American Journal of Applied Chemistry. 2015;**3**(6):207-211

[18] Iwata T, Ishikawa R, Ichino R, Okido M. Photocatalytic reduction of Cr(VI) on TiO film formed by anodizing. Surface and Coatings Technology. 2003;**169–170**:703-706

[19] Djellabi R, Ghorab RFM, Nouacer S, Smara A, Khireddine Q. Cr(VI) photocatalytic reduction under sunlight followed by Cr(III) extraction from TiO_2 surface. Materials Letters. 2016;**176**:106-109

[20] Chen G, Feng J, Wang W, Yin Y, Liu H. Photocatalytic removal of hexavalent chromium by newly designed and highly reductive TiO_2 nanocrystals. Water Research. 2017;**108**:383-390

[21] Lei XF, Xue XX, Yang H. Preparation and characterization of Ag-doped TiO_2 nanomaterials and their photocatalytic reduction of Cr(VI) under visible light. Applied Surface Science. 2014;**321**:396-403

[22] Ma CM, Shen YS, Lin PH. Photoreduction of Cr(VI) ions in aqueous solutions by UV/TiO_2 photocatalytic processes. Journal of Photoenergy. 2012;**2012**. Article ID 381971:7

[23] Wahyuni1 ET, Aprilita1 NH, Hatimah H, Wulandari AM, Mudasir M. Removal of toxic metal ions in water by photocatalytic method. American Chemical Science Journal. 2015;**5**(2):194-201

[24] Barakat MA, Chen YT, Huang CP. Removal of toxic cyanide and Cu(II) Ions from water by illuminated TiO_2 catalyst. Applied Catalysis B: Environmental. 2004;**53**:13-20

[25] Satyro S, Marotta R, Clarizia L, Somma ID, Vitiello G, Dezotti M, et al. Removal of EDDS and copper from waters by TiO2 photocatalysis under simulated UV–solar conditions. Chemical Engineering Journal. 2014;**251**:257-268

[26] Kanki T, Yoneda H, Sano N, Toyoda A, Nagai C. Photocatalytic reduction and deposition of metallic ions in aqueous phase. Chemical Engineering Journal. 2004;**97**:77-81

[27] Lenzi GG, Fávero CVB, Colpin LMS, Bernabe H, Baesso ML, Specchia S, et al. Photocatalytic reduction of Hg(II) on TiO_2 and Ag/TiO_2 prepared by the sol–gel and impregnation methods. Desalination. 2011;**270**:241-247

[28] Khalil LB, Rophael MW, Mourad WE. The removal of the toxic Hg(II) salts from water by photocatalysis. Applied Catalysis B: Environmental. 2001;**36**:125-130

[29] Li Z-J, Huang Z-W, Guo W-L, Wang L, Zheng L-R, Chai Z-F, et al. Enhanced photocatalytic removal of Uranium(VI) from aqueous solution by magnetic TiO_2/Fe_3O_4 and its graphene composite in environmental science & technology. Environmental Science & Technology. 2017;**51**(10):5666-5674

[30] Jiang X-H, Xing Q-J, Luo X-B, Li F, Zou J-P, Liu S-S, et al. Simultaneous photoreduction of Uranium(VI) and photooxidation of Arsenic(III) in aqueous solution gC_3N_4/TiC_2 over hetero structured catalysts under simulated sunlight irradiation. Applied Catalysis B: Environmental. 2018;**228**:29-38

[31] Aguado J, van Grieken R, López-Muñoz MJ, Marugán J. Removal of cyanides in wastewater by supported TiO2-based photocatalysts. Catalysis Today. 2002;**75**:95-102

[32] Turki A, Guillard C, Dappozze F, Ksibi Z, Berhault G, Kochkar H. Phenol photocatalytic degradation over anisotropic TiO_2 nanomaterials: Kinetic study, adsorption isotherms and formal mechanisms. Applied Catalysis B: Environmental. 2015;**163**:404-410

[33] Choquette-Labbé M, Shewa WA, Jerald A, Lalman JA, Shanmugam SR. Photocatalytic degradation of phenol and phenol derivatives using a nano-TiO_2 catalyst: Integrating quantitative and qualitative factors using response surface methodology. Watermark. 2014;**6**:1785-1806

[34] Nakano K, Obuchi E, Takagi S, Yamamoto R, Tanizaki T, Taketomi M, et al. Photocatalytic treatment of water containing dinitropheno L, and city water over TiO_2/SiO_2. Separation and Purification Technology. 2004;**34**:67-72

[35] Shukla SS, Dorris KL, Chikkaveeraiah BV. Photocatalytic degradation of 2,4-dinitrophenol. Journal of Hazardous Materials. 2009;**164**:310-314

[36] Khataeea AR, Kasiri MB. Photocatalytic degradation of organic dyes in the presence of nanostructured titanium dioxide: Influence of the chemical structure of dyes. Journal of Molecular Catalysis A: Chemical. 2010;**328**(1–2):8-26

[37] Vinu R, Akki SU, Madras G. Investigation of dye functional group on the photocatalytic degradation of dyes by nano-TiO_2. Journal of Hazardous Materials. 2010;**176**(1–3):765-773

[38] Dariani RS, Esmaeili A, Mortezaali A, Dehghanpou S. Photocatalytic reaction and degradation of methylene blue on TiO_2 nano-sized particles. Optik: International Journal for Light and Electron Optics. 2016;**127**:7143-7154

[39] Alahiane S, Sennaoui A, Sakr F, Qourzal S, Dinne M, Assabbane A. A study of the photocatalytic degradation of the textile dye Reactive Yellow 17 in aqueous solution by TiO2-coated non-woven fibres in a batch photoreactor. Journal of Materials and Environmental Sciences. 2017;**8**(10):3556-3563

[40] Wahyuni ET, Roto R, Sabrina M, Anggraini V, Leswana NF, Vionita AC. Photodegradation of detergent anionic surfactant in wastewater using UV/TiO_2/H_2O_2 and UV/Fe^{2+}/H_2O_2 processes. American Journal of Applied Chemistry. 2016;**4**(5):174-180

[41] Cappel CR, Cooley AC, Dagon TJ, Jenkins P, Robillard KA. Photographic Silver and the Environment. Rochester, New York: Eastman Kodak Company; 1992

[42] Dufficy TJ, Cappel CR, Summers SM. Silver discharge regulations questioned. Water Environment and Technology. 1993;**5**(4):52-56

[43] Concise International Chemical Assessment. Silver and Silver Compounds: Environmental Aspects. Geneva: WHO; 2002

[44] U.S. Environmental Protection Agency. Integrated Risk Information System (IRIS) on Chromium VI. Washington, DC: National Center for Environmental Assessment, Office of Research and Development; 1999

[45] Agency for Toxic Substances and Disease Registry (ATSDR). Toxicological Profile for Chromium. Atlanta, GA: U.S. Public Health Service, U.S. Department of Health and Human Services; 1998

[46] Cushnie GC. Electroplating Wastewater Pollution Control Technology. New Jersey: Noyes Publication; 1985. p. 375377

[47] Emmanuel J, Orris P. Mercury: Its Properties, Sources and Health Effects. United Nations Development Program Global Environment Facility Global Project on Health Care Waste; 2005. pp. 10-12

[48] Akpor OB, Ohiobor GO, Olaolu TD. Heavy metal pollutants in wastewater effluents: Sources, effects and remediation. Advances in Bioscience and Bioengineering. 2014;**2**(4): 37-43

[49] European Commission DG ENV. E3. 2002. Heavy metals in waste final report. Project Environment, COWI A/S, Denmark, Europe, 1–86

[50] Agency for Toxic Substances and Disease Registry (ATSDR). Toxicological Profile for Uranium (Update). Atlanta, GA: Public Health Service, U.S. Department of Health and Human Services; 1999

[51] Yoong LS, Chong FK, Dutt BK. Development of copper-doped TiO_2 photocatalyst for hydrogen production under visible light. Energy. 2009;**34**(10):1652-1661

[52] Gonell F, Puga AV, Lópeza B, García H, Corma A. Copper-doped titania photocatalysts for simultaneous reduction of CO_2 and production of H_2 from aqueous sulfide. Applied Catalysis B: Environmental. 2016;**180**:263-270

[53] Wang W, Zhang J, Chen F, He D, Anpo M. Preparation and photocatalytic properties of Fe^{3+}-doped Ag@TiO_2 core-shell nanoparticles. Journal of Colloid and Interface Science. 2008;**323**(43):182-186

[54] U.S. Environmental Protection Agency. Integrated Risk Information System (IRIS) on Uranium, Natural. Washington, DC: National Centre for Environmental Assessment, Office of Research and Development; 1999

[55] Bard AJ, Parsons R, Jordan J. Standard Potentials in Aqueous Solutions. New York: Marcel Dekker; 1985

[56] Razali MH, Fauzi AMN, Mohamed AR, Sreekantan S. Morphological, structural and optical properties study of transition metal ions doped TiO_2 nanotubes prepared by hydrothermal method. International Journal of Materials, Mechanics and Manufacturing. 2013;**1**(4)

[57] Siddhapara K, Shah DV. Study of photocatalytic activity and properties of transition metal ions doped nanocrystalline TiO_2 prepared by Sol-Gel method. Advances in Materials Science and Engineering. 2014;**2014**. Article ID 462198:4

[58] Khakpash N, Simchi A, Jafari T. Adsorption and solar light activity of transition-metal doped TiO2 nanoparticles as semiconductor photocatalyst. Journal of Materials Science: Materials in Electronics. 2012;**23**:659-667

[59] Al-Ahmed A. Metal doped TiO_2 photocatalysts for CO_2 photoreduction. Materials Science Forum. 2013;**757**:243-256

[60] Lo C-C, Hung C-H, Yuan C-S, Hung Y-L. Parameter effects and reaction pathways of photoreduction of CO_2 over TiO_2/SO_4^{2-} photocatalyst. Chinese Journal of Catalysis 2007; **28**(6):528-534

[61] Tan SS, Zou L, Hu E. Photocatalytic reduction of carbon dioxide into gaseous hydrocarbon using TiO2 pellets. Catalysis Today. 2006;**115**:269-273

[62] Akhter P, Hussain M, Saracco G, Russo N. Novel nanostructured-TiO_2 materials for the photocatalytic reduction of CO_2 greenhouse gas to hydrocarbons and syngas. Fuel. 2015; **149**:55-65

[63] Shehzadad N, Tahir M, Joharia K, Murugesana T, Hussai M. A critical review on TiO_2 based photocatalytic CO_2 reduction system: Strategies to improve efficiency. Journal of CO_2 Utilization. 2018;**26**:98-122

[64] Oluwafunmilola OM, Maroto-Valer M. Transition metal oxide based TiO_2 nanoparticles for visible light induced CO_2 photoreduction. Applied Catalysis A: General. 2015;**502**:114-121

[65] Abdullah H, Khan MR, Pudukudy M, Yaakob Z, Ismail NA. CeO_2-TiO_2 as a visible light active catalyst for the photoreduction of CO_2 to methanol. Journal of Rare Earths. 2015; **33**(11):1155

[66] Liu S, Zhao Z, Wang Z. Photocatalytic reduction of carbon dioxide using sol–gel derived titania-supported CoPc catalysts. Photochemical & Photobiological Sciences. 2007;**6**:695-700